THE EDGE OF REALITY

Two Scientists Evaluate What We Know of the
UFO Phenomenon

J. ALLEN HYNEK   JACQUES VALLÉE
Foreword by PAUL HYNEK

This edition first published in 2023 by MUFON, an imprint of
Red Wheel/Weiser, LLC
With offices at:
65 Parker Street, Suite 7
Newburyport, MA 01950
www.redwheelweiser.com

Copyright © 1975 by J. Allen Hynek and Jacques Vallée
Introduction to new edition copyright © 2023 by Jacques Vallée
Foreword copyright © 2023 by Paul Hynek

All rights reserved. No part of this publication may be reproduced or transmitted in any form or by any means, electronic or mechanical, including photocopying, recording, or by any information storage and retrieval system, without permission in writing from Red Wheel/Weiser, LLC. Reviewers may quote brief passages. Previously published by Henry Regnery Company, ISBN: 0-8092-8209-7. This new edition includes a new introduction and foreword.

ISBN: 978-1-59003-309-8

Library of Congress Cataloging-in-Publication Data

Names: Hynek, J. Allen (Joseph Allen), 1910-1986, author. | Vallee, Jacques, author.
Title: The edge of reality : two scientists evaluate what we know of the UFO phenomenon / J. Allen Hynek and Jacques Vallée ; foreword by Paul Hynek. Description: Newburyport : MUFON, Mutual UFO Network, 2023. | Includes index. | Summary: "Two eminent scientists debate UFO incidents and persuasive cases, and explore what we still need to know about the phenomenon"— Provided by publisher.
Identifiers: LCCN 2023027486 | ISBN 9781590033098 (trade paperback) | ISBN 9781633413221 (ebook)
Subjects: LCSH: Unidentified flying objects. | BISAC: BODY, MIND & SPIRIT / UFOs & Extraterrestrials | TECHNOLOGY & ENGINEERING / Aeronautics & Astronautics
Classification: LCC TL789 .H89 2023 | DDC 001.942—dc23/eng/20230629
LC record available at https://lccn.loc.gov/2023027486

Cover design by Sky Peck Design
Interior by Happenstance Type-O-Rama
Typeset in Adobe Caslon Pro, Montserrat, and Gill Sans

Printed in the United States of America
IBI
10 9 8 7 6 5 4 3 2 1

The inexorable alphabet inevitably places H before V; the resulting order of the authors' names is in no way to be construed as an order of importance. Although Vallée has done the yeoman effort in this work, he insists that the alphabet be honored.

—ALLEN HYNEK

CONTENTS

FOREWORD, BY PAUL HYNEK............................. vii
INTRODUCTION TO NEW EDITION, BY JACQUES VALLÉE........ 1
ORIGINAL INTRODUCTION 3

1 "I THOUGHT THEY WERE COMING TO GET US!"............ 11
The UFO Problem: A Statement in Six Points ❖ The Ely, Nevada, Case ❖ They're Playing Games with Us ❖ I Never Tell Stories ❖ Research on UFOs ❖ What Are the Basic Facts? ❖ Is It All Nonsense? ❖ Changing Viewpoints in Science ❖ The Investigation

2 "THEY JUST WAVED BACK AT US..."................. 47
The Sighting at Boianai ❖ The Second Night ❖ The Repeaters ❖ A Festival of Absurdities ❖ Hynek Goes to New Guinea ❖ The Progress of Research ❖ Radar Confirmations ❖ Psychic, Natural, or Technological?

3 THE SCIENTISTS AT WORK 63
Table of UFO Sightings by Astronauts ❖ A Jealous Phenomenon ❖ Why Aren't There Some Pieces? ❖ The Men in Black ❖ Who Should Study UFOs? ❖ Is There a Cover-Up? ❖ Three Avenues to a Solution ❖ A Computer Scenario ❖ The Analysis of Photographs ❖ Magnetic Effects

4 "IT WOULD BE TOO FRIGHTENING TO MANY PEOPLE"..... 83
Encounters with Occupants ❖ The Uses of Hypnosis ❖ "They Are Going to Catch Us!" ❖ The Pattern ❖ Betty Hill Comments on the Pascagoula Case ❖ The Problem of Contact ❖ Pascagoula and the Nature of Reality ❖ The Dangers of Hypnosis ❖ The Modern View of the Brain ❖ The Problem with Questionnaires ❖ Hynek's Hypnosis Experiment

5 THE NIGHT AN OCCUPANT WAS SHOT 107
"I'm Gonna Get It off My Mind!" ❖ "Have You Ever Shot at Night with a Scope?" ❖ The Loss of Time ❖ The Intimidation ❖ What Apparitions Do ❖ The Occult Connection ❖ What Children See ❖ The Landing at Eagle River ❖ How Fast Does Thought Travel? ❖ Are UFOs Hostile?

6 FLYING SAUCERS YOU HAVE SWALLOWED 137
An Air Force Hoax ❖ The Analysis ❖ Did the Star Capella Land on the Runway? ❖ The Caliber of the Witnesses ❖ The Life and Lies of George Adamski ❖ Dr. Hynek Meets "Professor" Adamski ❖ Is a Martian Buried in Texas? ❖ Planned Invasion Delayed

7 REMINISCENCES 153
The Invisible College ✧ Don't Rock the Boat! ✧ The Blue Book Fallacy ✧ There Couldn't Be Another Condon Report ✧ The Swamp Gas Incident ✧ Looking for the Haystack ✧ The European Scene ✧ The Frustration ✧ The Dangers of Cultism ✧ Behind the Scenes ✧ The Credibility Gap

8 THE LITERATURE 179
Personal Motivations ✧ Should We Burn the Condon Report? ✧ The Amateur Groups ✧ Some Intriguing Stories ✧ The Skeptics and the Damned ✧ The Function of Myth

9 BRAINSTORMING 191
Some Scenarios ✧ The Billionaire Friend ✧ Four Hypotheses ✧ The Nature of Contact ✧ Stretching the Imagination ✧ The Evolution of Man ✧ Secret Weapons ✧ Psychic Projection ✧ Interlocking Universes ✧ Conclusion

APPENDIX A A Sighting in Saskatchewan 215
APPENDIX B A Sighting in New Jersey 225
APPENDIX C UFO Sightings by Americans Have Doubled in Seven Years 233
APPENDIX D Soviet UFOs 235
APPENDIX E The Technological-Cultural Gap 237
APPENDIX F Scientific Discoveries as a Linear System 239
INDEX 240

FOREWORD

I suspect it may be fairly common for many people to learn about their parents not from their parents but from their parents' friends. My parents simply didn't start many conversations talking about themselves. I didn't know until well after their deaths that my father was an equestrian trick rider in college and that my mother initially posed as a buyer to help black families buy houses in the '60s.

Yes, I was there with my mom and dad in the 1970s as UFOs became such an integral part of the zeitgeist. But I was pretty young, and, like my four siblings—Scott, Roxane, Joel, and Ross—I didn't have to make any efforts to be involved in ufology; it was just there. I saw a lot of things happen, but I didn't know much about why.

In the past few years, I've come to learn so much I didn't know about my parents—from Jacques Vallée. In the course of many pleasant hours on stage and off with Jacques, I've learned not only interesting facts but also background, and what was actually going through their minds when I was talking to my parents back in the day.

Conversations about conversations.

It's from such conversations that I've had with Jacques, Paola Harris, and others that I got so many of the nuances behind why my father thought the way he did. While Jacques and I were together in Socorro, New Mexico, Jacques told me of the seminal role Lonnie Zamora's experience had on my father's comfort level in talking about alleged sightings of aliens.

My father was trained as an astrophysicist. He realized early on during Project Blue Book that he wasn't studying UFOs, he was studying UFO reports. And that instead of calculating the mathematics of orbital trajectories and making direct observations of stars himself, he was talking to other people about sightings of things they didn't understand that they had reported before he arrived, cases that included reports of objects flouting the laws of physics he knew so well.

My father wasn't trained in these soft skills. He became a traveling trauma psychologist to people who reported a whole different level of trauma, and who then most often faced good old-fashioned terrestrial stigma to boot. And I'm pretty sure the Air Force didn't provide him with that kind of on-the-job training.

So my father needed to turn to colleagues whom he could confide in, and whom he trusted. As he says in the book, even though Jim McDonald loudly disagreed with my father on working with the Air Force, the mere fact that McDonald took the subject seriously was in itself a relief.

My father talked to hundreds of illustrious people from top Air Force brass to Neil Armstrong to Dick Gregory. But his favorite talks were with engineers like William Powers, philosophers like Steven Gouge, and, most of all, with his good friend Jacques, who, more than most others, understood the challenges and benefits of data. It was in large part in conversations with Jacques like in this book that my dad explored some of the more esoteric aspects of the phenomenon. Jacques's keen interest in consciousness helped my rocket-scientist father relate some of the "Cheshire Cat" facets of the phenomenon to his underlying spiritual beliefs.

In *The Edge of Reality*, reading the conversations between Jacques, Arthur Hastings, and my dad, you get a ringside seat to some of the canon of ufology actually taking shape. Concepts like alien inoculation, control, and the meaning of the title. As individuals, we often seek ease in our comfort zone. But today's comfort zone is tomorrow's boredom, and today's adventure is tomorrow's comfort zone. And so it is for science—don't recoil from the "edge of reality," even if it isn't a certain path to academic glory. For just beyond it lies another science and knowledge.

I learned so much about my parents from conversations with their friends. Now with the conversations in this timely reprinting, we can all learn much about the deeper levels of this, and the next, reality.

PAUL HYNEK
LOS ANGELES, 2023

INTRODUCTION TO NEW EDITION

In today's race for science, given the complexity of hurried conferences, intense competition for funding, and the hassles of travel, it is rare for researchers to have the luxury of a few days away from the lab, in a friendly environment where they can put things in perspective and compare notes with just a few colleagues. Yet those moments are the most precious and also the most productive in science: You can describe experiences, experiments, and results; you can find the right moment to confess blunders, to express regrets for failed trials, to search for better tools; and you can learn from one another in shared confidence.

This book, compiled in 1975, is an example of such a dialogue, when Dr. J. Allen Hynek and I were finally able to put facts and ideas on the table and sort out hypotheses about UFOs over three days of recording, away from the campus, the media, and the demands of business. The meeting was made more special through the friendly, intimate "facilitation" of our dialogue provided by Dr. Arthur Hastings, a Northwestern-trained psychologist who had specialized in small-group communication at Stanford University. Arthur knew both of us well, and he was acquainted with the issues.

Professor Hynek is no longer with us (he died in 1986, ten years after these conversations), but his image is frequently on TV, his voice is still heard whenever the subject comes up, and his wisdom still inspires those of us who continue the work. This book is just one example of our many conversations; even today, it provides a useful guide to research on UFOs.

A visit from Allen to our home in California was always a happy occasion. My wife and I expected it with pleasure, and our kids were eager to show him their latest game or ask him about telescopes and planets. Then we reminisced about our time in Evanston in the early sixties, where I had worked between Dearborn Observatory and the computation center, compiling the first searchable database of the massive Air Force files under his guidance.

The mid-seventies were a very opportune time for our conversations, because the subject was quiet. The Condon committee set up by the Air Force in 1967 had concluded that UFOs were not interesting, so the media frenzy had died out. The *New York Times* and the Academy of Sciences had buried the inconvenient subject, ignoring the fact that the majority of the cases submitted to them remained unexplained. The "swamp gas" controversy of 1966, almost ten years in the past, had been largely forgotten (although incompletely forgiven, because the witnesses still felt betrayed). Even the sensational case in Socorro, which had taken place in 1964 when Allen and I had studied it together, was fading from memory.

We both felt free to launch a completely new phase of research, happy to see that witnesses all over the world were ignoring the neglect of the American Academies. Not only did UFO sightings continue, they were as dramatic and disturbing as before, and we were getting them through clean networks of dedicated individuals doing independent research. As the reader will see, major series of unsolved cases had buried the false conclusions of the 1967 Condon report. There was Pascagoula in 1973, and the strange case of Connersville. There was Saskatchewan and Woodcliff Lake in 1974, and many others. The details were fresh; the research was current in our minds. The reader will find them here, among important incidents unjustly forgotten.

If today's Congress really means it when it demands new research, this would be a nice place to start.

In 1975, when this book was compiled, history had already washed out the negative statements of pundits in science and those in journalism, although I will always remember Walter Cronkite, his back to a wall-sized photograph of the Andromeda nebula, reassuring the American public on prime time: science had not found any vicious UFOs out there, everybody should get back to business!

In a strange way, now that the Air Force had closed Project Blue Book, and the fake dramas on TV had vanished, we sometimes thought the best research opportunities were open. In the lighter exchanges recorded here, we felt both a heavy responsibility and an intriguing, tantalizing pleasure.

If the mid-seventies were a special time for ufology, they were an even more important period for Silicon Valley, where I worked with the early venture capitalists and futurists, eagerly learning from both, busy between the labs and Wall Street. The internet was emerging from the ashes of Arpanet, and a new culture of technology was demonstrating its prowess in major leaps of research, ignoring the naysayers and negating the pundits.

Allen loved the excitement of it, and was eager to build on our growing research. Two years later, Steven Spielberg would stun the world with *Close Encounters of the Third Kind*, using an expression Dr. Hynek had gifted to the English language.

He would be on the set, smoking his eternal pipe, happily watching the aliens who stepped down from the mothership. History had changed.

—JACQUES VALLÉE, 2023

ORIGINAL INTRODUCTION

We start with the incontrovertible fact that *reports* of Unidentified Flying Objects exist. Furthermore, such reports continue to be made daily, in all parts of the world, by all sorts of people, from all walks of life and of all degrees of education, training, and culture. The sheer number of such reports over the past quarter of a century or so is embarrassingly large—embarrassing because we still do not know what UFOs are. Some individuals believe they know. Many dismiss them all as nonsense, hoaxes, or the products of "overheated" imaginations. Others are sure that they represent visitors from outer space, or from another "dimension," or from other "timespace" systems.

The stark fact is that we do not know. What adds to our intellectual frustration is that there is ample evidence that for every UFO *report* made, at least a dozen UFO sightings go unreported, often because of fear of ridicule. "I didn't report this because I didn't want people to think I was nuts" is a statement familiar to all UFO investigators. Elementary statistical considerations indicate that the total number of "UFO events" or UFO experiences over the entire world must be staggeringly high.

The UFO represents an unknown but real phenomenon. Its implications are far-reaching and take us to the very edge of what we consider the known and real physical environment. Perhaps it signals the existence of a domain of nature as yet totally unexplored, in the same sense that a century ago nuclear energy was a domain of nature not only unexplored but totally unimagined and even, in fact, unimaginable in the framework of the science of that day. All this makes the study of UFO reports a tremendously fascinating subject, combining a sense of adventure into the unknown with the exhilaration of skirting the edge of reality and even the fear of what might be revealed beyond the edge of what we consider to be reality.

Now, one thing should be made clear at the start of this adventure. In fact, it is already clearly implied in the very term UFO. The U in UFO means simply unidentified, and nothing more. However, what is unidentified to one person or persons may certainly be identifiable by persons of greater technical training and experience. It has been demonstrated clearly that the great majority of what at first are reported to be UFOs are, after study by competent personnel, determined to be really IFOs, or Identifiable Flying Objects. It was shown by one of the authors as early as 1949 that some 80 percent of all initial UFO reports were in fact identifiable, and this, unfortunately, has deluded many into believing that somehow all UFOs could be shown to be simply IFOs.

Nothing could be further from the truth; those 20 percent of UFO reports that continue to elude identification are true UFOs, representing a phenomenon that is both as enormous in scope, and as puzzling, as any science has tried unsuccessfully to explain. In this book we are not talking about reports that arise from misidentifications of the planet Venus, or of high altitude balloons, or of meteors, or of aircraft, or of reflections in windows, etc., but about this large core of true UFOs. Over the years, so much research has been done on UFOs by organizations in this and other countries, especially France and England, and the case studies of true UFOs in the files of these organizations and of many private investigators are so numerous, that when one takes the trouble to look seriously at the total picture revealed by such studies, no question remains about the reality of the UFO phenomenon. "There is more to UFOs than fools the eye," might be a good paraphrase of the familiar adage.

One of the authors, J. Allen Hynek, was charged for years with the responsibility of trying to identify UFOs reported to the United States Air Force, and he succeeded in four-fifths of the cases. He approached the subject as a complete skeptic. It was he who in 1966 advanced the "swamp gas" theory for some of the famous sightings that occurred in Michigan. Therefore, one cannot accuse him of being a gullible believer in "little green men" or spacemen from distant solar systems. After careful reconsideration of his experiences as scientific consultant to the air force, during which time he studied thousands of reports and interrogated many hundreds of witnesses to UFO experiences, Hynek has come to consider UFOs as a real but unknown phenomenon.

Jacques Vallée came to the study of the UFO phenomenon through his investigations in France, after the great UFO wave of 1954 in that country. He carried out computer studies of the reports there and in other countries. He became intimately acquainted with the work of UFO investigators on the Continent and in Great Britain, notably Aimé Michel and Guerin in France, and Bowen, Creighton, and others on the staff of *FSR* (*Flying Saucer Review*) in England. He has maintained close contact with investigators in other countries as well, notably those in Spain and Brazil.

Vallée came to the United States in 1962, first to the University of Texas and later to Northwestern University, where he obtained his doctorate in computer science. While he was at Northwestern, he published two books on UFOs, *Anatomy of a Phenomenon* and *Challenge to Science*. His third book on the subject, *Passport to Magonia*, was written after he left Northwestern to join Stanford University. He has expressed particular interest in the possible paranormal aspects of the UFO phenomenon, especially those connected with the reported appearance of "UFO Occupants" or "Humanoids"; his third book contains a catalog of nearly one thousand reports of UFO landings, in one third of which UFO occupants have been described. He has made detailed studies of many of the best-reported instances of this particular phenomenon. There is probably no one in the world better acquainted with this most bizarre phase of the UFO phenomenon.

The association of the two authors has been a long one. It began at Northwestern, although they previously had had frequent correspondence. Hynek had become involved with the UFO problem in a considerably different way, and also somewhat earlier, than Vallée. It was while a professor of astronomy and director of McMillin Observatory at Ohio State University in 1948 that he was approached by representatives of the U.S. Air Force from the Wright-Patterson Air Force Base at Dayton, Ohio. He was asked to act as an astronomical consultant to Project Sign, the hastily organized project to evaluate the new phenomenon of flying saucer reports—was it a fad, a post-war craze, or was there more to it than met the eye? Was the phenomenon man-made and from some foreign country, or was its origin somewhere in outer space?

Answering these questions was the task assigned to Project Sign, and an astronomer clearly was needed to assist in the evaluation of reports which could arise from the mistaking of bright planets, meteors, etc., for "flying saucers." Reference to his final report to Project Sign reveals that Hynek attributed as many as one-third of the then current reports to astronomical causes, and another third to other natural phenomena, with some reports not amenable to evaluation because of lack of data. Some 20 percent, however, did not lend themselves to evaluation as the misreporting of natural events. Still, Hynek maintained strict skepticism, holding that even these might be solvable as the result of unusual circumstances, such as the observing of natural events under extremely unusual circumstances. It was not until much later, after being increasingly concerned about the total aggregate of puzzling, unresolvable reports from all over the world, that he began to believe that there was something most unusual going on, something that might possibly be beyond what is presently imaginable.

Project Sign became Project Grudge, and this eventually metamorphosed into Project Blue Book, following an impressive wave of flying saucer reports in 1952. (The term UFOs was not employed until Edward J. Ruppelt, the first director of Project Blue Book, coined the term.) It was at this time that Hynek was called back to active duty, so to speak. In a paper delivered at a meeting of the Optical Society of America in 1953, Hynek first hinted at a possible change of personal attitude, but he nonetheless maintained a public stance of "resident skeptic," stressing the large number of sighting reports that continued to be generated by misperceptions of natural objects. It was not until the last decade that he felt it his scientific obligation to call attention to a phenomenon, which might indeed turn out to be extremely important. This took the form of his only book on UFOs, *The UFO Experience: A Scientific Inquiry.* This book was written after (and considerably abetted by) his association with Vallée. Although now in different parts of the country, the authors have kept in close contact.

Now, as professional scientists and because of their long collective acquaintance with the subject, the authors have collaborated on a book that brings together their present thinking about UFOs. They hope it will prove of value not only to those already well acquainted with the UFO problem—for whom the book is primarily intended—but also for those who might like to know what the current status of the UFO problem is, what is being done about it, and what it is like to have been associated with the problem for so long.

For some of their joint working sessions on this book, the authors asked Arthur C. Hastings, a specialist in the psychology of communication, to join them. Dr. Hastings is also a researcher in parapsychology and has carried out first-hand investigations of poltergeists, telepathy, and other psychic phenomena. He and Vallée have frequently discussed the similarities of some UFO experiences to paranormal psychic events.

For the joint sessions, the authors asked Dr. Hastings to compile a list of questions on UFOs, and then to chair their discussions on those questions and the issues that arose from them. This allowed the authors to debate each other, follow up ideas in process, and to speculate more freely than they might otherwise have done. Several of these discussions are printed verbatim in this book, and reading them you can get a behind-the-scenes picture of how the authors grappled with many UFO problems in serious, and sometimes humorous, discussion.

What sorts of things do the authors think about UFOs? How frustrating is it to work so long on the problem and yet not have a positive solution? Is the UFO phenomenon really just a will-o-the-wisp—a grand illusion—or are we probing at the limits of reality? If we didn't think that the latter were true, neither of us would spend a further minute on the problem. Although this book is not a catalog of UFO cases, interesting UFO sightings, reported by quite sane people, are used abundantly to illustrate the bizarre and tantalizing nature of this utterly fascinating subject. All these strange and seemingly incredible reports, from all over the world—can everybody be either mistaken or crazy?

Isaac Asimov, doyen of writers on scientific subjects for the general public (and with whom author Hynek has had a number of friendly and provocative discussions), has fairly presented the problem of the UFO for the layman. "The trouble is that whatever the UFO phenomenon is, it comes and goes unexpectedly. There is no way of examining it systematically. It appears suddenly and accidentally, is partially seen, and is then more or less inaccurately reported. We remain dependent on occasional anecdotal accounts." Outside of the fact that the reports are not quite so "occasional," Asimov's summary is quite correct, but it merely serves to emphasize the difficulty of the problem, not to dismiss it. There have been situations like this in science before; meteorites were introduced to science only through a long series of anecdotal accounts, and were disbelieved for an equally long time. In medicine and psychology, many things now accepted started as unbelievable old wives' tales.

The authors are well aware of these difficulties and have often met to discuss them. Out of these meetings has arisen a deeper understanding of the complexity of the problem. Some of our recent discussions were taped, and we felt that no better way existed to present our views (not necessarily identical views either) than to preserve the format of controversy and thus, so to speak, invite the reader to share in our own gropings and puzzlements. Hence, a large portion of this book is in dialogue form, edited only to remove purely conversational idiom so as to present better the theme of the conversation.

The dialogue format illustrates, probably better than the typical book format could, the nature of the edge of reality we approach when we really plunge into the problem. We quickly see that there is no handy, "off the shelf," ready answer. "UFOs are someone else's spacecraft" is an appealing hypothesis, but a thorough acquaintance with the subject shows that, while this may turn out to be the answer, it is not an acceptable hypothesis as is. If UFOs are indeed somebody else's "nuts and bolts hardware," then we still must explain how such tangible hardware can change shape before our eyes, vanish in a Cheshire-cat manner (not even leaving a grin), seemingly melt away in front of us, or apparently "materialize" mysteriously before us without apparent detection by persons nearby or in neighboring towns. We must wonder, too, where UFOs are "hiding" when not manifesting themselves to human eyes. And, even though modern radars are highly mission oriented and so sophisticated that what they first see is filtered through a computer so that objects the radar isn't programmed for remain invisible, one wonders why, in spite of that, such tangible craft from elsewhere would not be very frequently picked up by radars less sophisticated and not so discriminating in what they choose to see.

If UFOs do turn out to be manifestations of intelligence from elsewhere, then their mode of manifesting themselves here on earth is by some method totally unknown—and perhaps even unimaginable—to us.

Another tempting "solution" is to conclude that *all* UFO reports are hoaxes (we have already eliminated the obvious identifiable reports, the IFOs). This has been tried, but unsuccessfully. And necessarily so, for it is a game of trying to "prove a negative."

What sort of UFO reports are we talking about? They are incredible events, generally inadequately reported, probably because of the shock of the occasion. Take, for example, this sketchy report from an outpost two degrees south of the Arctic

Circle. Attempts to obtain more information have been fruitless, because two of the principals do not even wish to be identified or to communicate. Perhaps this particular report, though made by the superintendent of the Northern Canadian Power Commission and involving an American professor "quite high up in one of your universities" and a young companion who also wishes anonymity, should be discarded, because it is of no scientific value without more data. We are, however, passing it on to you to illustrate the frequent frustration a UFO investigator feels when reports are grossly incomplete and the witnesses refuse to come to the fore. The incident was reported to us in a letter written by a French Canadian; for ease of reading, we have taken the liberty of correcting numerous misspellings in the original:

> Dear Sir:
> Recently I met a young man who stayed with us here in Chesterfield Inlet. Somehow we got on to the subject of flying objects and he asked me to tell you what three of us saw while out hunting. About two years ago the three of us were out hunting caribou. This American professor who is quite high up in one of your universities has asked us never to use his name in anything that we ever mention. However, I will get on with what we saw that has stopped me from ever going out to hunt again. And I mean this; I have gone through the war and never have I been so paralyzed with fear. And with a camera hanging from my neck and not able to take a picture!
> We had just turned around a bend in this ravine and came face to face with this object. The only way I can tell you what it looked like is to take two plates, one that is turning one way and the other one turning the other way, at such a speed that they seemed to glow orange and blue. The top had a dome and the under part had a slight dip in it.
> This object was no more than six feet from the ground when we first saw it. As you must be aware, in the Arctic we use ski-doos for transport, and one of the queer things was that when we came around this corner the three vehicles quit. The motors just quit. The noise that this thing gave off was a hum and yet not a hum.
> I know that I am not helping very much, but if anybody that comes in contact like we did and is still able to think straight, he is a better man than I am and I will admit it any time.
> On our return to Chesterfield Inlet, the professor chartered a plane and left, and has never returned to hunt again. Jerry, the other chap, quit the school here and left. I know where he is and I will try to get him to send you a letter. My job here in Chesterfield is Superintendent of N.C.P.C. I have been in the north for eighteen years. I don't really know whether I should send this letter; I don't want you to think I am ready for the funny farm either.
>
> Best regards,
> (name withheld)

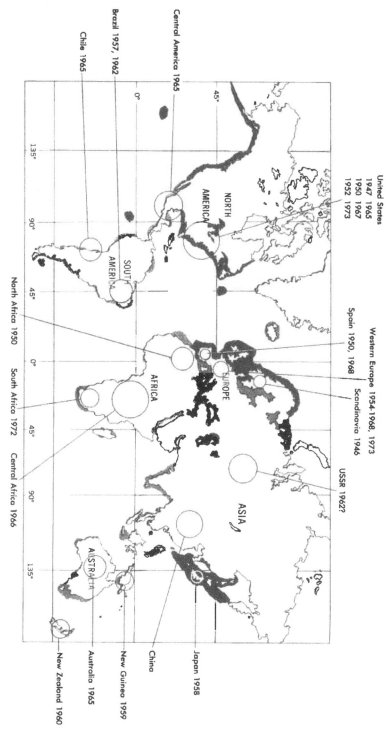

Figure 1 The global nature of the UFO phenomenon. (This map shows only the major waves of UFO events.)

How is one to regard such an incomplete report? A hoax? Always a possibility, and no way to disprove it without further data. Yet here is a responsible person writing a seemingly incredible tale from a remote northern region of Canada—a tale that could only cause him embarrassment if widely circulated. One must ask: For what purpose? Now, if that were the only tale like that, one could probably dismiss it, but UFO reports should be considered in their relation to each other and to the total pattern of reports.[1] Here is a report that fits a well-established pattern: the hovering a few feet above ground, the stopping of motor vehicles, the rotating disks, the hum, the glow. We believe such reports should be taken seriously. And we will now consider their implications to our scientific knowledge and to our awareness of reality.

---

[1] More detailed reports are printed here as Appendices A and B. Appendix B, in particular, vividly illustrates a struggle for comprehension at the "edge of reality." This letter from a seasoned newsman portrays superbly what happens when one suddenly confronts "the edge."

1
"I Thought They Were Coming to Get Us!"

An Excursion toward the Edge of Reality

"There is no hope of advance in science without a paradox."

—NIELS BOHR

The UFO Problem: A Statement in Six Points

The UFO phenomenon calls upon us to extend our imaginations as we never have before, to think things we have never dared think before—in short, to approach boldly the edge of our accepted reality and, by mentally battering at these forbidding boundaries, perhaps open up entirely new vistas. To many, such thinking is both frightening and a threat to their intellectual security. This book is clearly not for them. We want to explore with the reader the paradox of the UFO phenomenon, to imagine many things, to seek adventure in the unthinkable.

What sort of things will we be thinking about in the year 5000—and accept then matter of factly—that we would consider utter nonsense today? Undoubtedly Galileo and Newton would have considered the possibility of television to be nonsense, had they even been able to conjecture so far afield! Suppose, for instance, that by the year 5000 the interaction of mind with matter is far, far better understood than it is today. And suppose that it is commonplace then (although almost inconceivable today) that ideas can be projected through space at speeds faster than light and that such thought forms can be made to materialize and behave like material objects light years away from the "projector"!

Now this is something that we can *imagine*, even though we recognize it as sheer science fiction. But what about things we can't even imagine today? Of course there must be such things, because there have been such things in the past. Take, for instance, the fact that the sun is shining. One hundred years ago, it literally would have been unimaginable to imagine the correct explanation. At a time when it was not even known or imagined that atoms had nuclei, it would have been a little odd to think in terms of nuclear energy! Yet the sun shines by nuclear energy, which was not only not imagined but unimaginable a century ago.

It is indeed sobering, yet challenging, to consider that the entire UFO phenomenon might represent today what sunshine did a century ago—an unsolvable mystery—simply because we haven't explored that domain of nature yet. We must be aware of this possibility, but it should not cow us into hopeless inactivity. Research finally gave the answer to sunshine! The task ahead of us is to determine what about the UFO phenomenon are facts and what is fiction, then to take the facts and see what sort of hypothesis or hypotheses are needed (no matter how "far out" they may have to be) to explain the facts. If the facts demand that we look for extensions for our present knowledge of physics, then we must seek them, guided by intuition and observation. If the facts seem to demand a "paranormal" explanation, then let us boldly examine that avenue. After all, what one age regards as paranormal or supernatural is regarded as normal in another. The idea of radio would have been regarded as the work of the devil two centuries ago!

What, then, is the consensus among serious investigators of the UFO phenomenon? What facts have survived the sifting and analysis to the present? Where are we now? We may summarize our position as follows:

1. Truly unidentified sightings of events in the air or close to or on the ground exist.

To deny this would be tantamount to saying that everything that happens in the sky, in the air, and on the ground is fully understood. (We don't fully understand ball lightning, as just one example.) Reports of such sightings continue to be made by groups of credible and often technically trained persons from many parts of the world, despite repeated onslaughts of ridicule by the uninformed. Here is an incontrovertible fact: Such reports exist, and in far greater numbers than the public is aware. Fear of ridicule is a powerful deterrent for open reporting of such events.

Any notion that such UFO reports represent merely a fad, a popular craze, or that *all* such reports are the results of self-delusion and misperceptions of ordinary events simply will not wash. The evidence is all to the contrary when the UFO phenomenon is squarely faced "in the field" by personal investigation rather than in the armchair. Fads are short-lived; the UFO phenomenon is noted for its persistence over the years. And while the great majority of initial "raw" reports do find explanation as self-delusion, as every experienced UFO investigator can confirm, there remains a solid and mind-boggling residue of reports that defy rational explanation. It is most unfortunate that UFO reports made by the untutored who cannot recognize Venus, or bright meteors, or high altitude balloons continue to plague the UFO investigator. It is disturbing that much time must be spent in exposing such IFOs (Identifiable Flying Objects), but the more experienced the investigator, the more quickly such IFOs can be recognized and eliminated.

2. *The reported sightings that remain unexplained after serious examination fall into a relatively small number of fairly definite patterns of appearance and behavior.*

These patterns have been well delineated by UFO investigators and especially by the present authors.[1] The patterns have not changed during the past quarter of a century. There is little new in the contents of recent UFO reports; a 1975 report generally does not differ in basic content from a 1955 or 1965 UFO report. UFOs in a real sense do not constitute news. One would expect that if UFOs had no substance in fact but were entirely the products of human imagination, there would be considerably greater variety in UFO reports. It does not say much for human imagination to report the same old (but incredible) stories; over some three decades one would imagine that story tellers as well as out-and-out hoaxsters would bring some variety into their productions!

3. *The more educated and affluent the individuals, the more coherent and articulate is their UFO report, and the greater is their tendency to take the UFO problem seriously.*

Several national polls over the years have clearly indicated that "belief" in UFOs is correlated with education and income. If it were all a matter of credulity, superstition, and ignorance, one would certainly expect an inverse correlation! Similarly, it has been our experience that our most puzzling UFO reports come not from the untutored, superstitious, and gullible but from the "practical," sensible, the "I'm from Missouri" types, persons who in their daily lives have the reputation of being staid and stolid, no-nonsense persons. Here one must clearly and emphatically differentiate between the purveyors of IFO tales (Venus, balloon, aircraft, meteor-generated "UFO" reports) and the technically trained—from farmer to radar expert—who, when they can be persuaded to report, produce the truly puzzling cases, those reports which alone can be termed UFO reports.

4. *The UFO phenomenon is worldwide, and experienced investigators agree on the following basic reported characteristics:*

The content of reports made by persons considered most reliable, based upon the known conduct of their lives, describe apparently physical craft that can maneuver with ease in our atmosphere, appear to be largely unaffected by our gravity, and seem to abrogate at will the inertial properties of matter (as exhibited by the ability to hover effortlessly a few feet above the ground or high in the air and to attain incredible accelerations, generally noiselessly). They are detectable

---

[1] See Vallée's *Anatomy of a Phenomenon* (Regnery, 1965), *Challenge to Science* (Regnery, 1966), and *Passport to Magonia* (Regnery, 1974), and Hynek's *The UFO Experience: A Scientific Inquiry* (Regnery, 1972).

by radar on occasion, but not always, as attested by some of the best accounts that involve radar confirmations of visual sightings and vice versa. At night they are primarily visible by self-generated light and only secondarily by reflection, and virtually all colors of the spectrum are reported. Color changes are often observed as the craft accelerates.

The UFOs are capable of producing physical effects: they are reported to leave "landing marks" or other physical evidence of their proximity, such as rings or other types of imprints on the ground; plant life is often reported withered or blighted and sometimes scorched, and animals are often reported greatly disturbed by the presence of a UFO (it has often been reported that warning of the presence of a UFO was first given by the disturbed behavior of animals, especially dogs and cattle). Humans have reported physiological effects, such as temporary paralysis and blindness, headaches, and nausea, etc., but permanent or fatal damage has rarely been reported (and then not fully documented).

Inanimate matter is also affected by UFOs. Frequently, but not always, the proximity of a UFO is reported to interfere with radios, television sets, power lines, and especially with the electrical systems of automobiles. It has often been reported that car headlights, radios, and engines are temporarily put out of commission.

The question of whether the UFO phenomenon is a manifestation of some type of intelligence, whether extraterrestrial, "meta-terrestrial," or, indeed, some aspect of our own, is a critical one. Certainly the reported trajectories and general behavior of the UFO cannot be called random. There seems to be direction involved. And in those cases—and there are many—in which "occupants" or "UFOnauts," ostensibly the pilots of the craft, are reported, intelligent control of some sort seems obvious. Even if the occupants are robots, a more distant intelligence is implied. Another indication of intelligence (though to us it may seem perverse) is the almost universally reported response of these occupants to their detection: upon detection, they scurry away and take off in their craft. Except in a very few cases, there appears to be no desire for involvement with the human race (which of course might be regarded by some as the highest form of intelligent reaction).

The nonoccupant cases, which appear to be the majority, and which range from puzzling lights seen at night (whose behavior, general appearance, and trajectories do *not* conform to obvious explanation) to the metallic disks frequently reported in the daytime, to the domed, portholed craft reported mostly at night, all exhibit behavior which can be characterized as intelligent as contrasted to random-walk activities.

A very peculiar property of the UFO, and one which has caused many to dismiss the subject entirely, is the extreme localization of the phenomenon in space and time. "Why didn't more people see what so-and-so reported?" is frequently asked. The answer is very probably twofold: First, it has been the experience of most investigators that "close encounter" cases manifest preferentially in relatively isolated places, away from dwellings and installations frequented by humans (see Figure 2). This is evident from a study of specialized catalogs of these events from which as much "noise" as possible has been vetted. There seems almost to be some sort of an "avoidance principle" in operation.

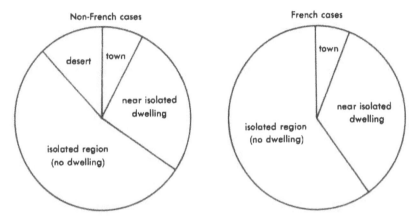

Figure 2 Distribution of Type-I cases as a function of population density. "Are they avoiding us?" UFO landings are a "jealous" phenomenon. Close encounters seem to take place in isolated areas. (Statistics by C. Poher.)

Second, UFOs are not seen by large groups of people, or sequentially by independent groups of people along the trajectory of a UFO, simply because vertical rather than horizontal trajectories are almost exclusively favored. Some skeptics (and heaven knows, we need them to keep a proper balance!) pointed out that when an extremely bright daylight meteor occurred, it was seen by large groups of people, photographed by many, and its trajectory accurately traced, but this has never happened with a UFO. But, this famous daylight fireball travelled on a slowly descending horizontal path, miles high, across several states, and it came at the height of the tourist season in the western states when the density of camera-equipped tourists is highest, whereas UFOs almost always are highly localized and favor the vertical path. Rarely has there been reported a UFO that travelled horizontally across several states! UFOs are most frequently reported as descending at a steep angle, hovering for a few moments, and then taking off again on a nearly vertical trajectory, and in a remarkably short time.

Another property of the UFO is the distribution of the times at which most are seen (see Figure 3). Vallée has shown (1963) that peaks in sighting times occur at 2200 hours (10 P.M.) and 0300 hours (3 A.M.). The ten o'clock peak may be due to the fact that more people are awake to experience a UFO (although one might think that a somewhat earlier time in the evening should be the case if this is the explanation), but the 3 A.M. peak (although much smaller than the 10 P.M. peak) cannot be so explained. In fact, if one rectifies the time distribution curve to allow for the fact that a much smaller number of people are available for a UFO sighting during the night, the true peak becomes overwhelmingly predominant (Figure 4).

All the above adds up to the fact that the UFO phenomenon represents a set of entirely new and empirical observations that our present scientific framework is severely strained to encompass.

5. *The UFO phenomenon has been ignored or very imperfectly studied by the scientific fraternity.*

This, we believe, has had several causes. Mainly, the data have not been properly presented. When UFO reports are found only in tabloids, pulp magazines, and treated sensationally in the media, this is certainly sufficient in itself to direct scientific attention elsewhere. Given the all-too-frequent sponsorship of the subject by the credulous believers, cultists, pseudo-religious fanatics, plus the clear fact of the low signal-to-noise ratio, it is understandable that the scientific fraternity chooses to ignore the subject. In TV parlance, the subject has had a poor sponsor.

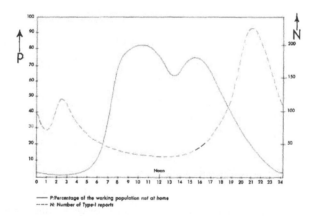

Figure 3 The observation that most UFO landings are reported in the evening hours does not tell us at what time the phenomenon itself is most frequent, because few potential witnesses are outside after 10:30 P.M. in most countries. Here the percentage of the working population *not at home* has been plotted on the same graph as the time distribution for 2,000 UFO landings.

"I Thought They Were Coming to Get Us!" 17

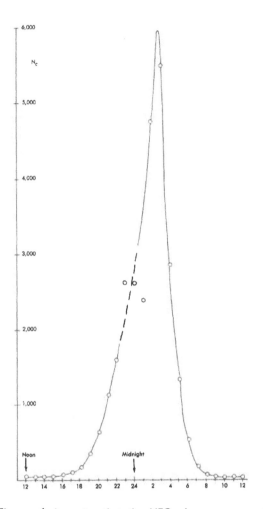

Figure 4 Assuming that the UFO phenomenon was independent of the presence of a human witness, Jacques Vallée has reconstructed the "time" distribution of the UFO landings as a function of the time of day. The maximum is now around 3:00 A.M., and the total number of cases is multiplied by a factor of 14. In other words, if people stayed outside at night, we would have some 28,000 reports of close encounters in our files rather than 2,000. In order to guess the actual number of landing events, one would still have to multiply by a factor representing the "reluctance to report," and by another factor representing population density, leading to a staggering number of events.

6. *The carefully gathered and sifted data are amenable to scientific study of an interdisciplinary nature but probably necessitating new departures in methodology.*

The lack of "hard" data is deplorable, but it is to a large degree ascribable to the gross lack of effort to investigate the phenomenon properly. The subject has never been given adequately funded, serious scientific attention. It is becoming clear, also, that the phenomenon is so strange that the methodology of investigation must be adapted to the phenomenon, and not the phenomenon to the methodology. There is a growing interest in, and an increasing open-mindedness about, the subject on the part of established scientists and of the educated public. An increasing number of knowledgeable persons no longer dismiss the subject as sheer nonsense and favor a serious investigation of its many bizarre aspects, sensing that this could well lead to new advances in science and general knowledge.

The above points represent the thinking of many UFO investigators, including, of course, the present authors, whose personal investigations of a great many cases support this consensus.

Because the need for serious study of UFOs and for making available authoritative and reliable information about the UFO phenomenon seems today greater than ever, a number of prominent scientists have created the Center for UFO Studies to meet three specific needs: first, to provide a central source for the gathering and dissemination of reliable information and technical reports about the UFO phenomenon; second, to provide a scientific body to which persons who have had a UFO experience can report without fear of ridicule and with the assurance that their reports will be of all possible use to science; and third—and most important of all—to provide a scientific body that will support the serious research into UFOs the subject demands.

Where might such studies lead? What can be studied, and how can it be studied? What is the *real* nature of the UFO phenomenon? Does it really originate from the actions of other intelligences in the universe? If so, where, and what, might they be? Does the UFO phenomenon have a purely physical explanation, or is there a vaster, hidden realm that holds the explanation? These are questions that have concerned the authors for many years, and to these questions we now turn our attention. We begin with a close-encounter case we have personally investigated.

The Ely, Nevada, Case

The following interview is from a taped discussion between an investigator and two brothers, Dave and Chuck (not their real names), who were the principals in what has come to be known as the Ely (for Ely, Nevada) case. The excerpt begins as Dave describes his initial reactions to the incident.

DAVE: I thought I didn't really see it . . . it took about thirty days for it to sink in my mind that I really saw that. The longer I went, the more real it got in my mind. I was trying to prove to myself that I really didn't see it. As time went on, I kind of convinced myself that I really did see it, because my brother was there also, and he saw the same things I saw.

INVESTIGATOR: Can you describe the whole experience as completely as possible?

DAVE: We left Idaho after we got our truck loaded with furniture and I drove. We had to run in shifts where I was sleeping and my brother Chuck was driving, and then I was driving and he was sleeping. I took the first shift. I guess we left the house about 12:30 or 1:00 A.M. . . . that was on February 14, 1974, a Thursday. I took the first shift and when we got to Wells, Nevada, he took the second shift after we gassed up, from Wells to Ely, which is about a hundred miles or so on Highway 93. I was sleeping in the cab of the truck and he reached over and shook me and said "Wake up!" He said he had seen some flying objects, and I told him he was crazy. Of course, I was tired; we had been up all night. I proceeded to go back to sleep and was probably asleep ten or fifteen minutes and again he shook me and said, "Hurry up, wake up, fast, quick." That's when I thought maybe I'd better find out what he was talking about. *Then I saw the first object.* Chuck told me to look out the left-hand side of the truck, and I saw that round orange object to the left of the cab. I know I saw it . . . it was there . . . I saw it and he saw it. Right after I sighted the first object on the left of the truck, I glanced out to the right, and that's when I spotted the three other objects that were . . . I couldn't estimate how high they were up, but they were up quite a ways and I observed them for a few minutes. One was just sort of going off and then coming on. The other two were a little higher and they were constant. Just about that time the truck, I think, if I can recall correctly . . . I don't know for sure . . . but I think we came around a curve, and as we hit this curve it felt to me like a blast of wind hit the back of the truck and it felt like it just picked it up and we were like floating . . . I was sitting there in the right-hand side of the seat watching him and he just kept going like this with the truck and he couldn't steer it . . . he didn't have any control over the truck![2]

---

[2] Far from being the only instance of loss of vehicle control, this observation fits an old and well-established pattern. On one occasion, a Canadian couple from Gleichen, Alberta, saw a beam of light directed at their car and perceived that the car was no longer on the ground. The vehicle was suspended about two feet above the highway and "flew" for a quarter of a mile at nearly forty-five miles per hour before settling down again as the "beam" was extinguished. On September 22, 1971, a typewriter mechanic from Brazil, Mr. Paulo Caetano, was driving near Bananeiras when a flying object came within thirty feet of his car, which skidded toward the shoulder of the road and stopped as the engine died (*Flying Saucer Review*, Special Issue No. 5, November 1973).

They're Playing Games with Us

CHUCK: I first saw this other object on the left side of the truck, on the driver's side. And it was about the size of a volleyball, I would say, and it was very bright in color. It was too large to be a star and it was very low—it was just highly illuminated. And at this time, I tried to wake my brother up, you know, to get his opinion on it. Dave was tired and said, "You're crazy, I want to sleep." So I didn't shake him too much—I didn't try to press it and get him awake. So, I guess about five miles down the road, one of the objects on the right side of the truck—one of the bluish-green objects—started to go faster than the others, and it went down about a mile and a half away from the others and it crossed over the road. The next thing I knew, it was over to the left of the truck and turned into an orange, luminous ball, and it was very low to the ground. So, I shook my brother. I said, "Get up! I want you to see what's going on here. What is this?"

He got up and he looked out and observed this orange object. He looked around, and I said that over to the other side of the truck there's some green-blue lights. At about that time, I noticed the light go out—the orange light—it just disappeared. And at this same time, when we were talking about it, I felt two blasts of air that came from the right side of the truck—just two gushes of air. It felt like I was losing control of the truck. The only thing I can compare it with is if you take a curve or something on ice, how the truck sways from one side to the other. . . . This is how it felt to me. But at the same time, the lights were flickering off and on, the motor was missing, and I was losing power. I know I was doing fifty-five miles an hour at the time this happened. I kept my foot on the gas, and I was losing power. Now, I want to tell you that what it felt like to me was that I was actually floating. And I was steering the truck, I was moving the wheel back and forth. I didn't have any control over the truck and we just coasted to a stop. So, when the truck finally stopped, I told Dave to get out of the truck and just look around, see what happened. Actually, I thought something had happened mechanically to the truck.

Dave got out with a flashlight, he looked under the truck, and he hollered back in and he said, "The driveshaft's turning." I said, "Well, that's sure funny because the truck's in neutral." He got back in the truck and about that time, I happened to look up and I saw down the road in front of us a light that covered the whole road—just a big, big, bright, luminous light. So I said to Dave, "What in the hell is that?" That's exactly what I said to him. And he said, "Oh, oh," he said, "they're playing games with us." I said, "What do you mean?" He said, "They're playing games with us. I know what that is up in the sky." And he said, "I know what that is up in the road." And then I started to get scared. And we sat there for a while and watched this object on the road. I forgot all about what was around us as far as the other bluish-green lights that I had seen, and I can't recall whether the other one, to the left side of the truck, that was up above the mountain was still there. I don't know. I forgot about it. All I was concerned with was what was on the road. So, Dave grabs the light and he gets out of the truck and he gets out in front of the truck and he shines the light at this thing. And to me it appeared like it was starting to move. So I hollered over to him, I said, "Dave, get back in the truck!"

INVESTIGATOR: Which one are you referring to now? The one in the road?

CHUCK: The one right out in the middle of the road, in front of the truck.

INVESTIGATOR: How far away was that?

CHUCK: I'd say probably 150 yards; you know, it's hard to determine distances at night. I don't know for sure, but I'd say it was about that close.

INVESTIGATOR: Actually on the road?

CHUCK: Right on the road, in the middle of the road. And it covered the whole road. So, when it looked like it was starting to move toward me, I hollered to Dave to get back in the truck; so he got back in, and I was scared then and I locked the door. That was the first thought that I had to lock the door. So when he got back in we both looked at it and we started talking. And, the way it appeared to me was that we were looking at a real bright light, you see what appears to be spears. You know, if you're looking at a star or something, you can see the reflection, I guess it is, that looks like spears coming off. And it had an outline of a reddish light on top. And we watched this thing for a while and then it just seemed as though—I don't know—it disappeared. Something came by us real fast.

I Never Tell Stories

INVESTIGATOR: Did anything else happen in connection with the sighting?

DAVE: I did have one experience with my mother. Prior to that happening, all I could think of was getting home. We had to get back to work, we had no time to waste. When we got to McGill, we called my mother and told her only that the truck had broken down; we told her we were in McGill trying to locate another truck, and to drive toward McGill and watch for us in each town. We didn't know whether we were going to get a truck or if we were going to have to stay in McGill, so I told her: as you come down the highway, just check, just keep checking until you catch up with us. They were still at home. We called from McGill after everything was over, some people picked us up and took us into the town around 6:15 or 6:30, I think. When we got to Barstow, we gassed up and started to leave and my parents pulled into the station. We got out of the truck and went up to them. The first words my mother said were, "My God, I had one heck of an experience last night." I said, what was it? She said, "You were in my room and I was awakened by your voice saying *Mom*, just a flat out, vivid *Mom*." And I said, do you remember what time that was? "It was around 4:15 to 4:30 in the morning." And I said, do you want to know something funny? And I told her what had happened! She didn't see my figure or anything, she just was awakened by my voice. She said she felt like I was there, like I was walking into her room but she didn't see me. It was just a verbal word. Nothing like this has ever happened before in our family. I've never had any ESP experiences or anything like that before. I've read about it and I believe it's true that these things happen, but I've never really come to an experience like this.

I didn't feel like I was getting any communication from them, I just felt like they had control of us and that they could do what they wanted with us.

INVESTIGATOR: What is your feeling now about the whole experience?

DAVE: I still sincerely doubted myself, I really did for a long time and maybe, I can't explain it, maybe it was real, maybe it wasn't . . . I don't have the answer and that's what bothers me, because I know what I saw. I don't think there's much the government can do about it. You read stories about things and all of a sudden it's quieted down and put on a shelf somewhere. I know there've been airline pilots who saw them, jet pilots, I've read accounts . . . and those are outstanding people . . . an airline pilot has to know what he's doing to fly an aircraft! I don't think the government knows anything about it unless maybe they're messing with it. I've had that feeling for a long time. Maybe they have something here we don't know about. My brother and I discussed it from the time we left with the new truck, all the way back, and we've tried to go back through ourselves to try and make each other believe or correlate what we saw. We kept going back over and over and over it to get the facts out.

Perhaps there are tests that can be run to see if the oil was drying up at a normal rate or whether it was a chemical reaction which would normally be produced . . . but what I can't figure out is why this drive shaft was spinning, and I can't get a mechanical answer to it. The man who took our truck said on the phone that the drive shaft had been welded into the rear end or twisted off and the wheels fell off when they raised the truck. . . .

I never tell stories . . . I sometimes don't believe it myself . . . I believe it one day and the next day I say, well, maybe it didn't happen. I hope I don't see any more and if I do, I hope I have a camera. If I'd have known that would happen I would have taken a camera with me.

Research on UFOs

In publishing the Ely case, can we expect it to be understood by those who may have only a superficial acquaintance with the subject? How can we convey the complexity and the relevance of such reports? This was the theme of a recent discussion. Dr. Arthur

Hastings, who served as moderator, got our debate underway by asking: "How can the subject of UFOs and the state of UFO research be presented to the public? Where should we start?"

VALLÉE: What I find in discussing the subject is that a tremendous amount of background work needs to be built up. But first, we need a statement, a no-nonsense, concentrated approach to what is known now, what has already been done, what needs to be done again, what has been established, *what we know*. Allen has been working on this for twenty-five years, and I've been working on it for some fifteen years. There are many things we know now that we didn't know fifteen or twenty years ago. Many hypotheses that had come up have now been rejected. The public should know about that, so the same questions don't come up again and again and again; for example, are UFOs secret weapons from the Russians? That kind of thing. We have to put these questions to rest. Then we have to show how large the accumulated amount of material is and what the state of research is now! One man doing this study in France, and this other man researching this other aspect in Spain; this one doing this in Great Britain, and this is what we know about South America, and so on—we will bring something from the U.S. research and how we approach it, and mention the Condon Report.[3] We will point to the existing literature so that after our presentation the reader is really up to operating temperature. Then we can start speculating. We can go back and point out that the sightings are full of psychological and social implications, and some are hoaxes. People are somewhat reliable as witnesses, but not as reliable as you'd think. Then, in the end, the real problem for us as scientists is, how far do we want to go in open, public speculation? One idea is to really build up an airtight case for the status of UFOs as a scientific problem. Refraining from speculating too much so that we don't convey the vague feeling that science is simply puzzled—so what? Science is puzzled all the time about lots of things! Pulsars, genes, differential equations, quanta. Two scientists come up and write a book, and the book says that science is puzzled, so what? People don't need that kind of thing any more.

HYNEK: I would start by pointing out that we don't study UFOs; we really study UFO reports, the data. I would start from some incontrovertible facts that not even someone like Menzel[4] can deny. UFO reports exist, that's a fact. Well, why *should* they exist? They *do* exist and they didn't exist at least in this sort of profusion

---

[3] For discussions of this controversial report, see Chapters 7, 8, and 9.

[4] Professor Donald Menzel, a retired Harvard astronomer, has been a critic of UFO investigation, arguing that the entire phenomenon has a trivial natural explanation.

in the heyday of the science-fiction stories when you would have expected them to. They do exist now. They come from all parts of the world! They are definitely global and they fit definite patterns. *One of the facts we now know about UFOs is what they are not:*

1. They are not all imagination, because if they were, we would be having reports of all sorts of things. Instead, we have consistent reports of disc-shaped lights and of craft-like objects.

2. We know that the reports we are getting at the Center do not come from kooks and crackpots. They come from people judged by all common standards to be sane; from people whose testimony would be accepted in a court of law in any other context.

3. We also know that it is not just a question of people seeing what they want to see; UFOs are not projections and imagination because, on the contrary, the better reporter always tries first to find a solution of his own. He tries to fit it into a rational picture; he is not the sort of person who jumps and says that light must be a spaceship. So, in general, I will try to knock down some of these popular misconceptions.

We also need to point out the facts that do not hit the papers, and we must emphasize that there are such things as close encounters. So often people seem to think that UFOs are always seen in the distance. They are not aware of the fact that there are bona fide close-encounter cases. And when they do talk about close encounters, they say, "Well, yes, but that was a guy that said he took a ride in a flying saucer and went to Venus." They don't understand that there are cases reported by sane witnesses, in which tree branches are found broken, marks were left on the ground, and they don't understand the magnitude of it. Take, for example, this case from the files of the Center for UFO Studies:

> On October 15, 1973, at least thirty workers saw a spherical object come down and *land among some trees* on an AVCO facility near Connersville, Indiana. No sound was heard.
>
> "It looked like a perfect sphere with a band of white lights running around the middle. I didn't count the lights but I know there were quite a few. . . . At the bottom of the sphere was what looked like a triangle of black light glowing a sort of bluish color. The lights did not flicker or dim or brighten, they stayed constant."

It always comes as a surprise when I say that a researcher named Ted Phillips has over eight hundred cases with recorded marks on the ground.[5] They are not

---

[5] *Physical Traces Associated with UFO Sightings*, compiled by Ted Phillips, Jr., published by the Center for UFO Studies, P.O. Box 11, Northfield, Ill. 60093.

aware that animals are affected, that people have had reactions such as burns and nausea after a sighting; also, the phenomenon of weightlessness and levitation seems to be reported time and again.

VALLÉE: Isn't it true that we can approach this from two major points of view? Listening to what you were saying, I see first the *human* point of view: society faced with rumors, stories, reports that describe the behavior of a witness, not necessarily biased, but a report from a human witness. You can study this as human behavior. Or you can take another point of view—that of looking at *the behavior of the UFO*. There are patterns in both categories. From the human point of view you can say—

1. It is happening to people in all countries and climates.

2. It apparently happens in cycles.

3. It affects physical life. As you said, it breaks branches, it leaves imprints, it produces marks, it affects electric circuits, it does all those things that we can observe physically. It's interfacing with us in our physical universe, in our *normal* surroundings and *normal* background.

From the point of view of the UFO now, there are certain other things that are true. These objects behave, they have certain characteristic kinds of behavior that have not been observed previously in physical objects—

1. They do not behave like aerodynamic things. They go through motions that *seem* to violate the laws of aerodynamics as we know them.

2. However, they behave consistently; the pattern of flight is consistent in all parts of the world.

3. They do not seem to be affected by the witnesses or by the sociology or psychology of the observers.

4. In the close approaches to the ground, too, there are certain characteristic behaviors. For example, the close approaches are not simply correlated with population density, as we might expect. So it is as if we were trying to study a race of beings with characteristic patterns of behavior that have never been observed before. For example, there are consistent differences between large objects from which smaller objects come out, and the objects that are seen in single flight through the atmosphere.

What Are the Basic Facts?

HASTINGS: It seems that a question we could ask here is, "Very interesting. What are those characteristics?" Which would be like saying, "What can you say about sightings that are close to the ground? What about the ones in the air?" Tell me a bit about those kinds of characteristics. I didn't know what you just said: I just thought all flying saucers were the same. What categories of characteristics can you enumerate? What evidence can you cite?

HYNEK: What you are asking is: What invariants filter through the human observer?

VALLÉE: Well, we can talk about the shape of the objects, their diameter, and the time of day when they are observed. The graphs in Figure 5 were done by summarizing a number of statistical studies I conducted over ten years ago with a computerized catalog.[6] If it were true that more UFOs are seen when people are outdoors, then the maximum number of reports would take place during the day. What we seem to observe is that, at least in the case of landings, the reports are most numerous in the late evening and before dawn.

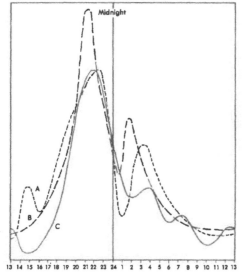

A: 362 cases prior to 1963, all countries
B: 375 cases in 1963-1970, all countries
C: 100 cases from Spain and Portugal

Figure 5 Frequency of Type-I reports as a function of time of day. In all countries and cultures, Jacques Vallée has found that the number of reported close encounters with UFOs is maximum in the late evening hours, typically about 10:30 P.M., with another peak just before dawn. Here the three waves correspond to cases reported before 1963, and cases from Spain and Portugal only. The three curves are almost identical.

---

[6] Based on the book *Challenge to Science* by J. and J. Vallée (Regnery, 1966).

Other categories that have been investigated are: the object, the reported size, angular diameter, distances from the observer, sound, the color, the luminosity, speed, the trajectory, noise, the duration of the sightings,[7] the choice of landing places, the number of witnesses involved and their ages, and the number of contact points with the ground. Some examples of these studies are illustrated in Figures 6 through 12.

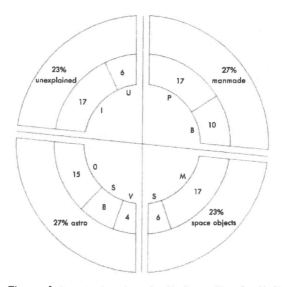

Figure 6 Statistics based on the Air Force (Blue Book) files. Between 1963 and 1967, Jacques Vallée computerized a catalog of all cases reported to the U.S. Air Force and reexamined the "official" explanations. He found that roughly one-quarter of the cases could be explained by manmade objects in the atmosphere (mainly planes and balloons). Another quarter was composed of space objects near the Earth: meteors and satellites. A third quarter contained astronomical causes: Venus, stars, or other objects. Fully one-quarter was unidentified, being classified as "unknown" or "insufficient information." This more complete analysis fully supports the analysis made by Hynek in 1949, in his first Project Sign report, in which roughly one-third of the then relatively few reports were ascribable to astronomical phenomena, one-third to manmade objects, one-sixth to "insufficient information," and one-fifth, or 20 percent, to truly unexplained phenomena.

---

[7] It is not true that UFO sightings are always of brief duration. On September 9, 1973, a television crew and police officers saw a bright unidentified object near Manchester, Georgia. They saw two objects in the sky, one of which faded away, but they observed the other one until dawn. Over the forty or fifty minutes of their observations, the light did not move, although the stars changed position.

Some other categories have been studied by Dave Saunders, and Ted Phillips has done research on the traces that were left by UFOs. Claude Poher in France and James McCampbell in the U.S. have done similar studies. We can talk about the time of the landing, the time of day, and the consistency of that data in all cultures, in all parts of the world. We can talk about occupants. Figure 5 shows the distribution of time of day for all the known close-encounter reports.

HASTINGS: Going back over the material, let me ask the question in a different way and see if it would change how you would answer it. If I said to you as scientists, put aside your concern for staying with the simple description of behavior. What would you say as scientists now, given normal criteria of evidence? What do you know about UFOs now? What can you say that has reasonably high probability?

VALLÉE: We know that there is an unknown phenomenon being manifested. It appears to center on a technological device, a machine that is capable of transporting occupants. The behavior of both the machine and the occupants appears to be consistent with the idea that we are faced with an alien form of life. However, their behavior is not consistent either with what you would expect from space visitors, or with what we know about physics. That's the dilemma.

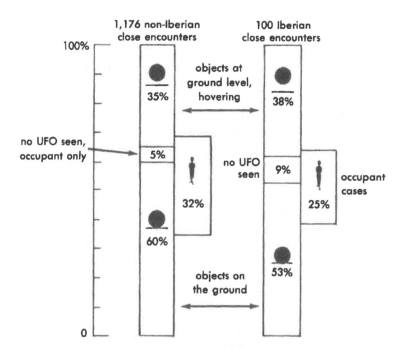

Figure 7 UFO invariants: The ratio of "landings" and close encounters to total cases appears to be similar in all parts of the world. Roughly a third of the close encounters involve "occupants."

"I Thought They Were Coming to Get Us!" 29

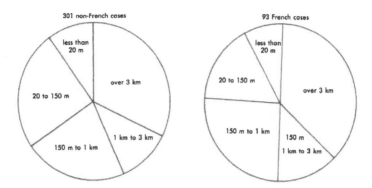

Figure 8 Distances from the UFO in 394 cases. These statistics by Claude Poher are in good agreement with those of Jacques Vallée and Ted Phillips. In half of the cases, the witnesses were able to examine the object from less than half a mile, and in many cases they were considerably closer.

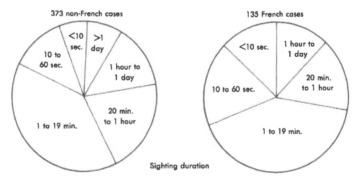

Figure 9 Sighting duration. The majority of UFO sightings last longer than a minute, typically five to twenty minutes (Poher 1973).

	Time Period						
Number of witnesses	Before 1947	1947 to 1953	1954	1955 to 1962	1963 to 1968	Total	%
1	19	39	118	83	232	491	56.0
2	6	10	49	38	77	180	20.5
3	4	6	12	13	22	57	6.5
4	1		7	12	10	30	3.3
5	2		2	1	5	10	1.1
6			1	3	3	7	0.8
7	1			1		2	0.2
8	1		1	2	1	5	0.6
"several"	7	10	19	13	14	63	7.2
"many"	11		5	9	8	33	3.8
Total	52	65	214	175	372	878	100.0

Figure 10 Number of witnesses in 873 close encounters. A significant number of UFO landings have several witnesses.

30 THE EDGE OF REALITY

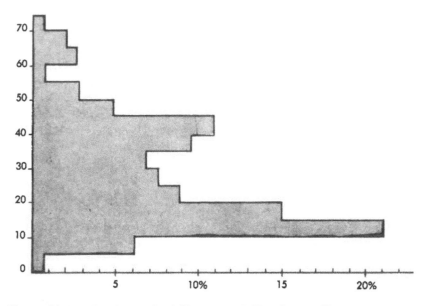

Figure 11 Age distribution for 147 witnesses in Type-I cases. Close encounters are reported by witnesses of all ages, although young adults seem reluctant to share their experiences for fear of ridicule.

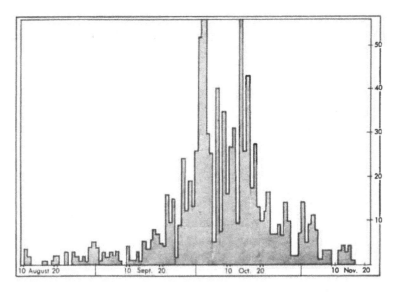

Figure 12 A typical UFO "wave." Number of observations made daily in the fall of 1954. From Jacques and Janine Vallée, *Challenge to Science* (Regnery, 1966).

We would be willing to accept something like an antigravity machine, if we had strange creatures coming out of it with a message for the president. We would even accept an antigravity machine causing psychic effects if its arrival were preceded by radio contact with a fixed source in space picked up by our radio telescopes, if it were preceded by five or ten years of exchange of data in the form of radio signals. But it is difficult for the human mind to accept either an antigravity machine or a technology that produces psychic effects that we don't understand, *when contact is not sought, or does not follow any of the guidelines science fiction has provided for our imagination.* Also, the UFOs seem to be a "crime without a motive." They don't have any purpose in coming here. They don't take anything. Apparently, they don't pose a threat. They do not bring help. They do not seem to affect our society on a collective level.

Is It All Nonsense?

HYNEK: When you describe a sighting like the Ely case, a person could say, "Well, you just proved that it is all nonsense." I was sitting next to a guy on a plane the other day and the subject came up; he was of the *it's all nonsense* school, so I asked him how many cases he had studied. "Of course, none at all," was his answer; "I wouldn't waste my time with that stuff!" In politics, religion, women, and UFOs, every man considers himself an expert. Everyone will tell you what's wrong with just about any subject, but they won't give any attention to a person who has studied that subject! However, if they were talking about brain surgery or heart transplants, they would acknowledge that there are experts who have studied it. In the UFO field, they simply feel that they know just as much about it as anyone else! So it's important to set the record straight. I can't add to what Jacques said. It's right down the line.

VALLÉE: Do we have to give a day in court to the man who believes that it's all nonsense? He may be a vanishing species. The latest Gallup report[8] shows that the more educated people are better informed on UFOs and have a higher ratio of belief in their reality. This percentage is steadily climbing every year.

HYNEK: Hell! One could spend all his energy confronting skeptics. That same energy is much better spent investigating the subject. Why waste time on people who have not bothered to learn the basic facts? It's their problem!

HASTINGS: A really crucial question: given what Jacques says, that the UFO has characteristics consistent with a technologically oriented vehicle and yet doesn't behave in terms of ordinary physics or in terms of motivational behavior, and

---

[8] See Appendix C.

given Allen's comment that people think it's nonsense, and that a lot of it does seem to be nonsense, how do you, as professional scientists, deal with *the necessity of keeping an open mind?* How do you make judgments and decisions, even though you can't make any final conclusions? If you could say that a particular case is clearly extraterrestrial, it would be no problem, you would have a solid conclusion. But you can't say that. So what do you do?

VALLÉE: First, it should be said that scientists are constantly faced with that kind of predicament. For example, there have been mysteries in the physics of the solar corona that have appeared almost as "magical" and irrational as anything UFOs do. The sudden manifestation of millions of tons of mass in a very highly excited state far away from the sun where there is no obvious mechanism to get it there was a major mystery for physics ten years ago; it isn't anymore. It has now been explained, but scientists are faced constantly with such mysteries, especially in astronomy where they have to deal with very large-scale, fundamentally puzzling phenomena. Not to mention the rotation of galaxies and things like that! The image people have of science is of a place where there are no mysteries. "Scientists know the answers." But in reality that's not the case at all. A scientist is someone who is permanently immersed in the unknown. That's why he is a scientist! Now the different thing about UFOs is that the public is interested in them in a major way, and the *public is the source of the report of the phenomenon itself.* The scientist has a very small probability of observing a UFO, and if he makes an observation, the likelihood of that being a close-encounter case, with lots of information bits, is very low.[9] There is little he can do to improve that. So he has to rely on the public. He has to do a public relations job to get the data. There is no fundamental problem dealing with something which is a mystery—that's what science is all about! But in most cases, those mysteries fall in the province of the scientists who keep it

---

[9] There have, however, been a number of instances in which UFOs were observed by astronomers, amateur astronomers, and Moonwatch teams. A recent case took place on Sunday, October 21, 1973. The president and vice-president of the Astronomical Society of Long Island (Lee Gugliotto and James Paciello), both associated with Vanderbilt Planetarium, while looking for meteors, noticed a reddish light that appeared to be coming toward them. Soon it had become brighter than Venus and had a diameter close to a third of the full moon. The two men called their wives and observed the thing as it moved from left to right, quite slowly, and rose above them in the sky. Then the white glow disappeared and was replaced by three blinking lights: green, white, and red. They were blinking randomly and were not spaced as the lights of an airplane would be. The four witnesses attempted to follow the object in their car but lost track of it. The late Otto Struve reported a "Nocturnal light" type of UFO to Dr. Hynek. Recently strange lights were sighted by astronomers at the Observatory at Mauna Kea, Hawaii, and a prominent radio-astronomer reported a light that maneuvered in apparently intelligent fashion.

to themselves until they can explain them. Most people have never known the sun had a corona, and it doesn't bother them that it behaves mysteriously! But they do know about UFOs.

HASTINGS: So we have, in effect, public pressure that demands an explanation. How does that affect your viewing this situation? People obviously are waiting for you or Allen to say, "At last we now have the solutions, and everybody can breathe easily!" Or words to that effect.

VALLÉE: We can say, "At last there are serious scientists, competent scientists, investigating this openly."

HYNEK: We know damn well that the UFO problem is not all nonsense. The Menzel school is untenable. There are certain things you can say about the UFO. That you can't explain it as a Russian vehicle. They just don't have that sort of technology. There is no such technology. You know that you can't explain the Betty Hill incident as a misinterpretation of Venus (see Chapter 5). We have separated out the things it isn't.

VALLÉE: There is another question, though. Should the public be turning to us for this kind of answer or should they be turning to the intelligence community, to the military, even to priests? Who are we to give that kind of answer? Is it our view here that science has the answer?

HYNEK: Well, they are turning to us because we supposedly have been with it the longest. We have the responsibility of saying something that we ourselves feel. They will expect an answer, and they will suddenly be disappointed if we speak in terms of probabilities. Suppose we said there is a probability of 6 percent they come from Alpha Centauri and 29 percent that they're from YB Virginis; is that the kind of an answer the public wants?

HASTINGS: There's a psychological need for an answer.

VALLÉE: We have to be careful with labels and people who want an answer *at all costs*. The medical people know this well. "It's a virus going around." We have to examine the concept of "naming" as doctors do, which tends to relieve anxiety when they don't know what causes certain symptoms. "It's the Hong Kong strain!" A new label is not a new answer!

HASTINGS: The UFO Virus.

Changing Viewpoints in Science

VALLÉE: Suppose we took a different angle. Suppose we said: "This is where science is stuck, *and these are the obstacles that need to be removed in order to find an answer.*" Given the state of science today, there is clearly no answer. If the speed of light can't be exceeded; if time can't be reversed; if psychic phenomena cannot be triggered by objective means, then there is no answer in that framework. So, *we have to look for another framework.* If we find that other framework, then who cares about UFOs? The effect will be much greater than just explaining UFOs! If that is true, then we should try to get some material on what it's like to be on the other side of the speed of light, and what it's like to have time reversal and causality violations. There's a lot happening in physics now. All sorts of people running around pretty close to Nobel Prizes with very, very, "way out" ideas about what theoretical physics should go into.

HASTINGS: Could you mention some of those ideas?

VALLÉE: Well, the fact that you have a mathematically consistent universe on the other side of the speed of light is a case in point. It is true that if you approach the speed of light from this side, you could never quite reach it, because your mass would get infinite. On the other hand, photons are going at the speed of light, aren't they? They suddenly got there! How did they get there? Look at the equations and assume you don't go *through* the speed of light but you *are* beyond that speed. Assume that you go at twice the speed of light, or three times the speed of light. Then plug this into the equation and see what happens; you find that you have a perfectly consistent universe. It's just this discontinuity between the two.

HYNEK: I want to hit this point very hard—to have people realize what it would mean to shift the axis and go back to the covered wagon days or the colonial days and have a little fantasy here. Go back to Ben Franklin's day and state the questions that were asked then but couldn't be answered and which do have an answer today. There were tremendous mysteries: how the sun shines, or how fast it is possible to go. We can think of many such things that couldn't be answered in Franklin's time because there weren't enough data, enough physics. If somebody had said that the sun got its energy by eating up its matter, that would have been regarded as magic. Benjamin Franklin couldn't have guessed that the sun was sort of a nuclear reactor. In that same sense, we may be unprepared, or science is, at the present time to solve the UFO problem. Ben Franklin could have made certain assumptions: he could have thought that if matter is not indestructible, then one could explain how the sun shines. But the conservation of matter was an absolutely sacrosanct thing, so the thought never occurred to him, or to anyone else. You would have been driven out of any scientific society if you even suggested that matter was not conserved.

VALLÉE: I am very impressed by one thing that you have said before, Allen, that the UFO problem was a good opportunity to teach the public how science worked. This really hasn't been done. The Air Force, the Condon Committee, had that opportunity, but they didn't use it. I think the fact that the public turns to the scientist for a definitive answer illustrates one flaw in the way scientific education is done. To the student, a scientist is someone who has written books on something, one who *has the answers*. We don't teach research, we teach the answers. Many are led to believe that knowing science is really regarded as knowing one's way through a library! The idea we give to students is that when you have a question, you go to someone who knows a little bit more; thus a PhD is someone who knows the library much better than someone who has a master's degree! If the PhD doesn't know the answer, then you go to the library again yourself and you search for the magic book that has the answer. If the book doesn't have the answer, then you hire a scientific consultant, and you ask him for the answer. But on the real frontiers of research, there are no answers! People can't accept that. I think it's a matter of education.

HASTINGS: It occurs to me that if the public were as excited about quasars or as worried, or as curious about quasars as they are about UFOs, they would be putting tremendous pressure on science to get busy on quasars. That, of course, isn't the case. A quasar is somewhat remote from people's everyday life—it doesn't visit people in the countryside! But UFOs interact with people's everyday lives. Secondly, UFOs don't seem to be natural phenomena, or science would be perfectly happy to investigate them. I think there's an unmentioned premise here that says that they aren't natural because they are not seen as natural in the descriptions and in the sightings; therefore, science is hesitant to investigate. If scientists thought they were some weird form of meteorite, if they had all the characteristics of natural phenomena, then there would be no resistance. But this is a step beyond. Even when the French Academy said that meteorites didn't fall from the skies, they didn't deny that they existed, that something happened, or that they were rocks or something. They just denied that they fell from the sky.

HYNEK: They created some sort of a cockeyed solution. They said they were rocks hit by lightning. They were of a mind to force things into the known framework. When Russell wrote his book on the origin of the solar system (a problem that has not been solved to this day, incidentally), he spent several chapters on the characteristics of the solar system. He pointed out that any theory of the origin of the solar system must explain the following things: the planets are in a plane, the heavier or more massive planets are farther out, they all revolve in the same direction. Their moons, with a few exceptions, revolve in the same direction. Natural satellites are all essentially in equatorial planes of the planets except for our moon. The planets have the greater share of angular momentum in the system. He set down all the different things that any viable theory would have to explain. Here's something that we should do here;

here are the things that are testable. The properties we have listed earlier. The fact that UFOs stop cars. There could be quite a few references. All the definite things we can say, after twenty-five years of research.

There are the things we can say with reasonable certainty that the UFOs do. Now, can science explain it? Obviously, it cannot. But neither could Ben Franklin explain why the sun shines. All right, what would have been necessary, what would we have had to do to science in Ben Franklin's day to have encompassed why the sun shines? That's exactly our problem. *What do we have to do with science today to make it capable of explaining this?* Science may have to be torn apart at the seams, even as classical physics had to be abandoned and quantum mechanics brought in to deal with the physics of atoms.

VALLÉE: But you can't always do that. For example, in technology forecasts, you find that there is always an element of serendipity which cannot be foreseen; at the end of World War II, there was a commission to study future weapons; they extrapolated all technological fields to imagine what the future weapon would be. They knew that H-bombs would come along—this was in 1946–47. So you had H-bombs, and you had projections in terms of rockets, you had projections in terms of weapons, the size and weight. However, no one predicted nuclear missiles, because they didn't consider electronic miniaturization! It was totally due to the invention of the transistor, which no one had predicted! So that weapon just wasn't predicted. The strategic planning for the next twenty years was just wrong on that one account, that nobody thought of the transistor.

HYNEK: Let me again give you an example: the great Lord Kelvin once wrote a very snotty article called "The Theory of Uniformitarianism Briefly Refuted." In that article, he gave the geologists a hard time. The geologists were finding fossils, and they needed a longer time frame than the physicists were giving them to explain them. If fossils were living here two hundred million years ago, then the sun had to be in existence two hundred million years ago! Kelvin said, "I'll give you three million years and not a day longer!" He said this because he used only the scientific reasoning of his day. At that time, they didn't have the slightest idea of nuclear energy, so the sun had to get its energy by contracting, and therefore the sun had to be relatively young. Secondly, he had measurements on the manner in which heat was flowing from inside the earth. He said, "Even if the earth had been completely molten in the beginning, in so many million years it would cool down to the present rate of outflow of heat." But he didn't make allowance for an unknown factor. In the case of the heat, he didn't make allowance for the existence of radioactive materials inside the Earth. It turns out that these are producing the heat! Of course, he refused to allow that there might be something he didn't know—in this case it turned out to be nuclear energy. He was wearing blinders. Although he was a topnotch physicist, he simply couldn't allow that science didn't know everything. And so he gave the geologists three million years and not a day longer! In fact we now have good evidence that the Earth is over four *billion* years old!

VALLÉE: What do we have to change in science to make UFOs possible?

HYNEK: The scientists!

HASTINGS: What specific reactions have you observed that might help people see how some scientists react to this? There was the man who said to you, for example, "It's nonsense, of course, I haven't looked at anything!"

HYNEK: It is the same reaction that medical men had toward acupuncture ten or fifteen years ago. Can you imagine someone giving a paper on acupuncture? Or continental drift? UFOs aren't so strange, after all.

VALLÉE: That doesn't mean to say that anything goes and that you can dream up any kind of magic because eventually science will explain it.

HYNEK: Anything you dream up has to fulfill a certain list of conditions. Of course, if you want to say, as a religious person, that it's all God's will, or if you want to adopt an *ad hoc* theory that is so broad that it explains everything, well, then that's no good.

HASTINGS: You can go at it from one way. Obviously, Kelvin and a lot of people were simply taking assumptions that were presently held and then inferring conclusions from them. Now the obvious thing would be to take what you, as scientists, believe to be true about UFOs and simply infer what necessarily follows from that information.

HYNEK: That's what I was saying. We have to infer that the physical world, as *we* know it, is not the sum total of our environment. And I think we have to admit the possibility that there is something from the paranormal world present, because of some of the things that UFOs exhibit, some of their properties. Time and again a cloud develops around it, and the object either changes shape or completely "dematerializes." There's nothing in present-day science which allows for dematerialization.

The Investigation

In May, 1974, the two authors made arrangements to meet with the witnesses in the Ely case and record an interview with them. It is published here with their permission:

VALLÉE: In the transcript of the previous interview, Dave said, "They are playing games with us." Chuck asked, "What do you mean?" and Dave said, "They are playing games with us. I know what that is up in the road." Now, I was intrigued by that. What made you say that? Did you suddenly have the feeling that you had some reason for knowing what it was? What made you say that? You were driving and you woke him up. And you said, "Those things are out in the road," and he said, "I know what is out there in the road."

CHUCK: Yeah, what happened is that he said it when we actually noticed the one on the road after the truck had stopped. Then he made that statement.

DAVE: It appeared to me they were coming after us. That is what I thought. That's why I think I said it. I mean, it seemed to me they were playing games.

VALLÉE: As if you had some insight into what they were after?

DAVE: No, not that they were after anything. It was just felt they were there, or somebody was there. Whatever it was.

CHUCK: Isn't it strange that you would make that statement, because it is the furthest thing from your mind, then all of a sudden these things start happening. And you pop out with a statement like that.

HYNEK: That's right, those things are interesting. It is funny how striking at one point of thought sometimes suddenly brings something totally unexpected to the surface. Things you were not very aware of at the time.

DAVE: I figured by that time that maybe there was really something there that was doing something—after all the events that had happened.

VALLÉE: Did it put you in mind of other incidents that you knew about? Something like what had happened to other people?

DAVE: I guess I had probably read certain happenings. You read things like that in the newspapers and it could have been that. I don't know for sure, but I never thought that would happen to me, or that I would actually experience something like that, really.

VALLÉE: Had you ever driven a truck like that one before?

CHUCK: Yes. I used to drive a truck. I drove one for four years at night.

VALLÉE: Of the same type?

CHUCK: Basically of the same type.

VALLÉE: So you know how it felt—taking a curve at a certain speed, and so on?

CHUCK: Right. Basically the same type truck—maybe a little bit longer—but you know the truck was the same.

VALLÉE: So you would know if it was something mechanical that was wrong, and you would know how to react?

CHUCK: If it was something mechanical, from what I hear really happened to the truck when they picked it up, *I just can't understand why I didn't crash or lose control of the truck.*[10]

HYNEK: The truck was behaving completely normally when you first started the trip? There was nothing unusual about it?

CHUCK: No noise of any sort. We had just gassed up prior to this. And everything just started happening and then it rolled to a stop. And it just doesn't seem like everything would be going wrong with it at the same time. If it was just the rear end—*why would the lights go off? And why would you lose power?*[11] You know, like the engine was missing.

HYNEK: Check me if there is anything wrong in my memory: You gassed up and were driving along and Dave was asleep. The first thing you noticed was one light on the left and three on the right?

CHUCK: All together four bluish-green lights.

HYNEK: Were they lights like stars or were they discs?

CHUCK: Well, from what I recall they appeared to be just like round, bluish-green lights and they were right in a row, one right after another.

HYNEK: Right in a row?

CHUCK: Right. I saw no flickering or anything. It was just a colored light.

HYNEK: Vertical or horizontal?

CHUCK: Horizontal.

HYNEK: And the sky was quite clear at the time? You said. . . .

CHUCK: It sure was.

HYNEK: I have forgotten whether there was a moon out. Was there a moon out?

CHUCK: I don't recall seeing one.

HYNEK: It was not like a star or like the moon?

CHUCK: No.

---

[10] Analysis of the parts of the truck by specialists at the Center for UFO Studies showed that the rear axle had broken on one side, not an unusual occurrence if the bearings had been subjected to a lot of friction and had gradually been worn out. The coincidence of the breakage with the observation of the UFOs and the loss of electrical power in the truck is unexplained.

[11] Mechanical difficulty can explain the loss of power at this point. The flickering of the lights of the truck, however, cannot be attributed to this cause.

HYNEK: They were all the same color?

CHUCK: Yeah. They all looked the same.

HYNEK: Did they appear to be connected by some rigid rod or did they at all—did they give you the impression that they were a part of a solid object?

CHUCK: No. They looked like they were separate. Separate lights up there.

HYNEK: Dave was asleep. Now what time did you try to wake him up?

CHUCK: When I saw the bluish-green lights. I believe that was the time I tried to get him awake. To take a look, you know.

HYNEK: Shortly after? A minute or so? Right away?

CHUCK: I think I looked at them for a little while. When I noticed they were following along, I thought, wake him up!

HYNEK: Were they at your three o'clock position—your one o'clock position? Or how far to the right did you have to look to see them?

CHUCK: Well if I am sitting at the wheel, I am looking out like this.

HYNEK: So it's about twenty degrees elevation and about the two o'clock position. All right, so at this point you tried to raise Dave and Dave said, "To hell with it," or words to that effect.

CHUCK: Right. He was tired and asleep and thought I was more or less crazy or something. Anyway I ignored the fact and kept watching them. I was driving and then one of them accelerated.

HYNEK: The front one—rear one—or middle one?

CHUCK: The front one. It accelerated and took off—and I don't know the distance, probably a mile to a mile and a half down, it crossed out in front of the truck, over the road. And after it crossed it seemed to come back, or I caught up with it, and it turned into an orange-colored glow. And it was low.

VALLÉE: Did it start changing color when it accelerated?

CHUCK: No. It appeared to be exactly the same, and next thing I know it's an orange ball. Just an orange ball.

HYNEK: Were you going straight at this time or were you going around a curve?

CHUCK: I feel I was going straight. I can't recall going around a curve.

VALLÉE: Were your lights on high beam or low beam?

CHUCK: I don't believe I used the high beams on that truck at all. So I would say they were probably on low.

VALLÉE: And how many cars did you pass or how many cars did you see on that road? Say half an hour before and half an hour afterward?

CHUCK: I don't believe I passed any—to be honest with you. We didn't see any traffic at all. At this period of time we were out there by ourselves. And when I started seeing these lights, I can't recall any cars coming by passing me or anybody behind me.

VALLÉE: Can you identify any specific time when the object changed color? Or was it just that when the next time you looked it was a different color?

CHUCK: That was just about it. When it crossed over the road it was blue. Like I say, when it left and went out in front of the truck it still looked the same as the others and the next thing I know it was over on the side of me and what probably caught my eye then was the fact that it was a big orange ball. All of a sudden it was there. So I woke him up. I said, "Something's crazy here."

HYNEK: When did you have the first glimmer of the thing on the road ahead of you—when did you first see that?

CHUCK: After the truck had come to a stop—Dave got out of the truck, looked underneath, got back in—and it just seemed like you looked up and there it was, back on the road, covering the whole road!

HYNEK: And when did the first trouble with the truck happen? Was it after the orange object was near?

CHUCK: It seemed as though the problem with that truck—all these things started happening when that orange light seemed to disappear. It actually started as the object appeared. Something went wrong with the truck, and about that time the orange object was gone, disappeared. It went out like shutting a light off.

HYNEK: You actually saw it go when the light went off?

CHUCK: Yes, I was looking at it one moment and then looked again, and it just went dark.

VALLÉE: You were awake at that point, Dave? You were watching?

DAVE: The first thing I saw was the large ball outside the left side of the cab. That's when I got awake and was aware of what was going on. I saw that, and it was definitely there. I don't know what it was. It looked like—*like if you looked at the sun and it wasn't blinding you.* It was just a solid round orange ball.

HYNEK: How many feet off the ground, above the ground, would you estimate it to be at that time?

DAVE: I would say ten or fifteen feet.

VALLÉE: Did it throw off any kind of shadow?

DAVE: You see something that is pitch black, pitch black and there it is.

VALLÉE: Could you see any part of the landscape underneath? Did it illuminate the landscape?

HYNEK: It could not have been very bright if it was ten or fifteen feet from the ground and not illuminating the ground.

DAVE: I would say it was kind of dullish—a dull orange color—it wasn't bright, it was dull.

HYNEK: It seemed like it was physically real, there in front of you?

CHUCK: Oh yes. There was no doubt that light was out there in front of that truck and covered the whole road. It was something that was enormous in size. I don't feel it was a moving car coming at us, because this thing was sitting out there and it covered the whole road. And we sat there and looked at it for a while, and Dave got in front of the truck and shined that light at it. That's when I really got scared because I think it acted like it was going to come at us. All the time it was constant there; it wasn't moving. When Dave got out there and shined that light at it, that's when it looked like it was going to move toward us and I hollered to him, "Get in the truck! It's moving." And then I locked the door and I felt fear.

HYNEK: There was this light ahead on the road, and there were still three greenish lights off to the right, stationary?

CHUCK: Right.

VALLÉE: The engine was running?

CHUCK: It was. And the lights of the truck were on.

VALLÉE: Did you try to turn the high beam on to illuminate the thing?

CHUCK: I never even thought of that.

VALLÉE: Strange that you would think of taking the flashlight and not think of turning the high beams on.

CHUCK: Well, he had the flashlight in his hand because he had gotten out in front of the truck. Now I don't even know why he had even gotten out in front of the truck. Because I didn't have the feeling that I wanted to get out and approach it.

VALLÉE: Now when you got out, was there any noise? Any sound whatsoever you could hear?

DAVE: I don't recall any other than the drive shaft turning.

VALLÉE: There was no sound from the object?

DAVE: No. None at all.

VALLÉE: When you were sitting in the truck and you were looking at Dave, could you see his face illuminated by the light out there—were there any shadows of him?

CHUCK: I don't even recall that. I don't know whether I even looked at him. Once we were talking and we were looking at that thing, I don't know whether I looked over at him or what.

DAVE: You felt like you were trapped, in other words. Couldn't go back, couldn't go forward, the truck wouldn't move, you were just sitting there waiting for something to change. And then that object seemed to come toward us, and then all of a sudden something came by the truck.

CHUCK: It was what he said that really scared me, "They're playing games with us," and my mind started thinking that could be highly possible. That it could be something from outer space, or whatever you want to say.

VALLÉE: When that thing came by, you hadn't seen it coming?

CHUCK: Just at the time he shined that light out there it appeared to me that it started moving. He jumped back in, and I didn't actually see it moving up closer. All of a sudden it was there and something went by.

VALLÉE: Could you see it in the rearview mirror? Did it give off light from the back?

CHUCK: It was strictly from the front. It was there and went by real fast, and it was all over. I don't even remember if that object was up there after that point. I don't remember anything about lights; it just seemed like a relief—it's gone! It just seemed like whatever happened was over with.

HYNEK: The green lights, too?

CHUCK: Right, I don't even remember if they were there, because that thing had my attention, it was out on the road, and that's when I was afraid. When it was gone, I was relieved and forgot everything. The next thing I thought about was how were we going to get this truck off the road. We got out to put the reflectors out.

VALLÉE: This is the kind of thing that is so shocking that a person dreams about it afterwards. Have either of you had any dreams that seem peculiar?

CHUCK: I haven't had any dreams about it, but I would lay in bed at night the first week or so after it happened, and try to relive it, try to get an explanation for what happened. Did I really encounter some UFOs? This goes through your mind. Nothing happened to me. I'm the same right now as I was then, but I experienced this strange thing. There was no contact other than my seeing this stuff, so there was nothing to make me say definitely that was an object from outer space, because I didn't see anybody, I didn't get a chance to walk up and see a metal ship out there, so I forgot about it, more or less.

VALLÉE: What about you, Dave?

DAVE: No, I didn't have any dreams about it, but I did more or less think it over and over again in my mind and try to figure out what really happened. I guess I'll never know that.

CHUCK: Let me tell you one thing. For quite a while, even right now, I feel goose bumps when we dwell on it and think about it. I get this feeling of a goose bump effect.

HYNEK: Later on, when you got to Barstow, you were in the filling station and your folks drove in in back of you, and it was already night again. One of the most amazing things happened. I guess the books on psychical literature are full of things like that: under emotional stress, somebody quite far away who is attached to you has this experience. Which of you did your mother think it was, was it you, Dave?

DAVE: It was me.

HYNEK: Which would imply I guess that you were the most emotionally disturbed by it. And your mother said she didn't see you but she said your presence was there and you said "Mom" or something like that?

DAVE: And also my Dad said he was laying on a cot and the cot started shaking and he also woke up, so both of them woke up. She stated to me that she just heard me say "Mom" just as clear as a bell right in the middle of the room, and I was there. But she did say it was about 4:30 in the morning.

CHUCK: Believe me, she had no way of knowing, unless she stopped on the way down and talked to somebody that we may have talked to. I did not tell her on the telephone what we had seen on the road. She did know that we had broken down, but that was after it all had happened. I didn't relay to her at all anything until we got to Barstow. She did make a statement that she wasn't a bit surprised when we called her. She was expecting the call because she had woken up at 4:30 in the morning feeling Dave was in the room.

HYNEK: She must have felt that something had happened, probably even she thought that there was a very serious accident.

CHUCK: That's right. It might well have been, if we hadn't stopped we could probably have flipped and doing fifty-five miles an hour. . . .

2

"They Just Waved Back at Us..."

The Sightings in New Guinea and the Worldwide Efforts to Understand UFOs

The public expects a constant supply of exciting stories from the press and the media. Sightings of UFOs are caught in this cycle of social excitement; new cases appear, occasionally become front-page material overnight, and are dropped forever a few days later. Yet for those who do research in this area, some of the older cases remain just as exciting as the hot news of today. The sightings at Boianai, New Guinea, made in 1959, retain all their mystery after many investigations and many attempts at explanation. The case is exceptional in the quality of the witnesses, the number and clarity of the descriptions, and the "communication" it implies between human witnesses and the occupants of a UFO. Let us quote from the account published by Reverend Norman Cruttwell of one of the early sightings in the series:[1]

The Sighting at Boianai

Boianai is a village on a small tongue of land made by the Mase River where it flows out of a deep gorge of the Owen Stanleys. It is on the south side of Goodenough Bay, some twenty miles across from Menapi. About four miles behind it the mountains rise sheer to culminate in two peaks which overhang the gorge on either side, Mount Nuanua and Mount Pudi. They are about

---

[1] "Flying Saucers over Papua," Rev. Norman E. G. Cruttwell, *Flying Saucer Review*, Special Issue No. 4, August, 1971.

4,000 feet high. Behind them rise ridge upon ridge, up to Mount Simpson, nearly 10,000 feet, which caps the range.

Right on the beach is the Mission Station of All Saints, Boianai, with a coral cement church and various mission buildings. It faces northwards, the beach running northwest to southeast. It looks across to the low hills of Giwa and Menapi on the Cape Vogel Peninsula.

The Missionary-in-Charge, Rev. William Booth Gill, is an old friend of mine. He came out to Papua with me in 1946, and I know him very well. On April 9, 1959, he was on his little 16-foot launch about a mile off shore, coming home from visiting an outstation. It was 6:50 P.M., and just about dark. The weather was clear overhead, but there were clouds and rain squalls about. The mountains were a dark silhouette against the still glowing sky.

He suddenly noticed a bright white light "like a Tilley lamp," apparently high up on the flank of Mount Pudi, not far from the summit. He estimates that the light was about 500 feet from the top. It was quite stationary, and he immediately thought, "Oh, there must be someone up there with a Tilley lamp." The Papuans with him all noticed the light. He was puzzled about the light, but not unduly so, and looked away, continuing to read his book. Five minutes later he looked up again, but the mountain was in darkness. The light had disappeared. This again seemed odd, but he took no notice, and went on reading. After another five minutes he was aware of the light again, shining out from the mountainside, but to his surprise it was shining from a completely new position on the opposite side of the mountain. It had moved a good mile to the east, quite impossible if a man had been carrying it.

However, Father Gill did not realize the significance of what he had seen, and looked away again. Next time he looked back, the light had gone, and did not reappear. The next morning he examined the mountain by daylight, and realized that there was no house or village or even any track up there, but only the precipitous mountainside. It was not until he got a letter from me about the later sighting from Giwa that it occurred to him that it might have been a UFO.

Father Gill was again a witness in a later observation that took place on June 26, 1959. This time there were thirty-eight witnesses in all, and their descriptions of the UFO and its occupants have become classic. Following is an extract from a report of the event by Rev. Cruttwell:

Father Gill had just had his dinner and came out of the front door of the Mission House. There is a small patch of lawn, a few trees, including coconut palms, and then a drop of perhaps fifty feet to the shingle beach below.

He casually glanced at the sky and looked for Venus, which was conspicuous at the time. In his own words: "I saw Venus, but I also saw this sparkling object which to me was peculiar because it sparkled, and because it was very, very bright, and it was above Venus and so that caused me to watch it for a while, then I saw it descend towards us."

Stephen Gill Moi [a native teacher in the mission], who joined Father Gill a few minutes later, described it as "shining with a bright white light, like a Tilley lamp." Ananias [a native worker] used exactly the same expression. Stephen remarks that it appeared to wax and wane in brightness, as though it were approaching and receding. Eventually it came quite close and hovered at a height which Father Gill estimated between 300 and 400 feet, though he admits that it was very hard to judge the altitude at night, not knowing the size of the object. He estimates its apparent diameter as about five inches at arm's length. Stephen said that if he put his hand out closed it would cover about half of it.

Father Gill states that it changed from a brilliant white light, when it was far off, to a dull yellow, or perhaps pale orange, when it was close. When asked whether he thought it was metallic, he answered:

"Well, it appeared solid, certainly not transparent or porous; we just assume that it was metallic from our experience of things that travel and carry men."

All witnesses agree that it was circular, that it had a wide base and a narrower upper deck, that it had a type of legs beneath it, that it produced at times a shaft of blue light which shone upwards into the sky at an angle of about forty-five degrees and that four "human figures" appeared on top.

Two of the witnesses state that they saw about four portholes or windows in the side, which they have indicated in their drawings. These are not indicated in Father Gill's drawing. Commenting on this he said: "I saw what appeared to be panels in the side of the object which glowed somewhat brighter than the rest, but I did not interpret them as portholes. I did not indicate them in my rough drawing."

Discussing the appearance of the object and its occupants, the report by Rev. Cruttwell goes on:

There was a certain discrepancy in the witnesses' estimate of the number of legs, though all agreed that they were in two groups. Father Gill is emphatic that there were four legs, tapering, two at each end, somewhat divergent.

I have reproduced Father Gill's drawing and the drawings of three of the witnesses for comparison (Figure 13) together with my own composite drawing

(Figure 14) which appears to embody the true appearance of the object from the description of the witnesses.

Here are Father Gill's comments on the "men":

"As we watched it [the object] men came out from this object, and appeared on the top of it, on what seemed to be a deck on top of the huge disc. There were four men in all, occasionally two, then one, then three, then four; we noted the various times the men appeared. And then later on all those witnesses who were quite sure that our records were right, and that they agreed with them, and saw the men at the same time as I did—were able to sign their names as witnesses of what we assume to be human activity or beings of some sort on the object itself.

"Another peculiar thing was this shaft of blue light, which emanated from what appeared to be the center of the deck. The men appeared to be illuminated not only by this light reflected on them, but also by a sort of glow which completely surrounded them as well as the craft. The glow did not touch them, but there appeared to be a little space between their outline and the light. I have tried to indicate this in the drawing. They seemed to be illuminated in two ways: (a) by reflected light, as men seen working high up on a building at night caught by the glare of an oxy-acetylene torch, and (b) by this curious halo which outlined them, following every contour of their figures and yet did not touch them. In fact they seemed to be illuminated themselves in the same way as the machine was."

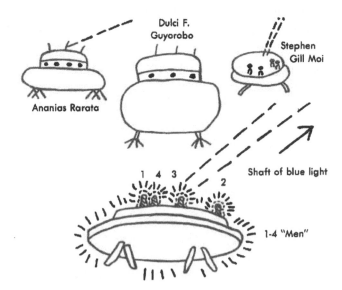

Figure 13 Boionai observations, June 26, 1959. Three witnesses' drawings and tracing of the drawing of the object with "men" by the Reverend W. B. Gill.

Figure 14 The Reverend Norman Cruttwell's charming, imaginative drawing of the scene at Boionai on June 27, 1959.

This is indicated in Father Gill's diagram.

When asked whether he thought they were wearing space suits he replied, "I couldn't say. It may be so; that would seem to be a possible explanation of the double outline, but I could not see any such suits."

I asked him whether he could see any details, such as the color of their skins. He replied that they were too far away to see such details, but that he would say they were probably pale. As for the details of their bodies, all he could be sure of was that they had the outline of normal human beings from the waist up. Their legs were hidden by the sides of the craft. If wearing clothes, they were very tight fitting.

The Second Night

The object came back the next evening and repeated its performance. An American astronomer, Dr. Menzel, has claimed that the object all the witnesses were excitedly watching was the planet Venus!

This hypothesis can be rejected on astronomical grounds, let alone others. On June 27, the sun set, local time, at 6:15 P.M.; hence the sky would still be quite bright at 6:02 P.M., when, according to Father Gill's notes, he was alerted to come see an object sufficiently prominent in the sky to cause consternation among the natives. Venus simply would not have been that prominent in a still-blue sky.

Reverend Cruttwell, who is familiar with the mountainous terrain at Boianai, independently confirms this:

> At six o'clock the sun would have only just gone behind the mountains, and the sky would have still been bright until half-past six. It would not have been really dark until at least 6:45 P.M. This rules out any possibility of the object having been a planet, such as Venus, which would not have been at all bright at such an early hour. Here is Father Gill's report of the happenings of Saturday, June 27:

> "A large UFO was first sighted by Annie Laurie Borewa, a Papuan medical assistant, in apparently the same position as last night. The time was about six o'clock."

> She called Father Gill, who came out at about 6:02 P.M., and saw the object for himself. It had the same appearance as last night's object, but seemed a little smaller, probably due to increased distance. Father Gill continues:

> "I called Ananias and several others, and we stood in the open to watch. Although the sun had set behind the mountain, it was quite light for the following fifteen minutes. *We watched figures appear on top—four of them—there is no doubt that they were human.* This is possibly the same object that I took to be the 'mother ship' last night. Two smaller UFOs were seen at the same time, stationary, one above the hills, west, and another overhead.

> "On the large one, two of the figures seemed to be doing something near the centre of the deck—they were occasionally bending over and raising their arms as though adjusting or 'setting up' something not visible. One figure seemed to be standing, looking down at us (a group of about a dozen)."

Father Gill, describing this later to Rev. Cruttwell, told him that the "man" was standing with his hands on the "rail" looking over, "just as one will look over the rails of a ship." Father Gill's report continues:

> "I stretched my arm above my head and waved. To our surprise *the figure did the same.* Ananias waved both arms over his head, then *the two outside figures did the same.* Ananias and self began waving our arms and all four seemed to wave back. There seemed to be no doubt that our movements were answered. All the mission boys made audible gasps (of either joy or surprise, perhaps both).

> "As dark was beginning to close in, I sent Kodawa for a torch and directed a series of long dashes towards the UFO. After a minute or two of this, the UFO apparently acknowledged by making several wavering motions back and forth (in a side-direction, like a pendulum).

"Waving by us was repeated, and this was followed by more flashes of the torch, then the UFO began slowly to become bigger, apparently coming in our direction. It ceased after perhaps half a minute and came no further.

"After a further two or three minutes the figures apparently lost interest in us, for they disappeared below deck.

"At 6:25 P.M. two figures reappeared to carry on with whatever they were doing before the interruption. The blue spotlight came on for a few seconds, twice in succession."

The Repeaters

The Father Gill case raises a new kind of problem. The UFO phenomenon appears to be worldwide in scope, appearing preferentially first in one area of the world and then in another. How can we explain the fact that some areas, at certain times, are specially "favored" by an abundance of sightings? This constituted a topic of recent discussion during a two-day "brainstorming" session at the home of Jacques Vallée, in which Dr. Arthur Hastings again participated.

HASTINGS: Is there any reliable sort of evidence that for some sightings, at least, the UFOs have displayed intentionality?

VALLÉE: There are situations that can be interpreted as intentionality on the part of the UFOs, especially in cases where they seem to manifest repeatedly with respect to persons and with respect to places. There are people and areas (like Boianai) that seem to be "repeaters." But, in some cases, there are also recurrent events, leaving ground traces, *even when there are no people around* to watch the objects themselves.

HYNEK: The repeaters and the "hot areas" are not just statistical fluctuation; it's way, way beyond that. That is, one naturally expects some bunching, some deviation from strict uniform distribution, but the observed repeating or bunching is definitely beyond the bounds of chance. And that's another fact we will have to explain.

HASTINGS: How could you analyze the New Guinea case and other close-encounter cases scientifically?

HYNEK: That's one of the questions: Is it a scientific problem? Can it be approached scientifically? Certainly the UFO itself can't be brought into the laboratory. True, photographs can be analyzed, ground samples can be analyzed, tree branches can be looked at in biological labs, and we can see whether or not we can grow things in the soil. Also, the radar traces and photos can be analyzed. To some limited extent the subject has been brought into the laboratory. There are certain aspects that can be brought in. Now one can say that if UFOs were a perfectly natural phenomenon, such as, let us say, the northern lights, or even ball lightning (which is pretty mysterious, but no physicist would regard it as anything but natural), it certainly could be studied scientifically. Then science has a certain approach: gathering data, analyzing it according to well-known principles, formulating some hypothesis about it. If, however, you are dealing with a phenomenon that is intelligently controlled, as the objects in New Guinea seemed to be, then the ordinary methods, the kitchen-tested methods of science don't quite apply, because you have a game-theory aspect coming into it. As Peter Ustinov said in that play, "You mean they know that we know that they know that we know?" It becomes an interplay of wills, and so you have to apply a somewhat different method. Now then, if it is that, one must be on the lookout for more complications. You have to look out for the fact that deception may be practiced. Is there any evidence that would lead us to think that it's a feint, that they do one thing but mean something else? There are some elements of this in the Father Gill case.

HASTINGS: Let's follow that up for a moment. What sort of things suggest that there's some kind of a deception or there's some gamesmanship going on?

A Festival of Absurdities

HYNEK: Well, all right. I'll have Jacques follow up on this, but one fact is apparent, what Aimé Michel calls "the Festival of Absurdities," the things seem to make so little sense to us! Why would they frighten animals, stop cars, douse headlights, unless it is a purely secondary effect and it wasn't intentional? Why did they just wave casually to Father Gill?

VALLÉE: Another example of this fact is that, in many cases of UFOs on the ground, the witness seems to be there precisely at the right moment when the UFO is on the road being repaired. There are many cases on record where people driving suddenly find a UFO on the road, with occupants pretending to be repairing it—which makes no sense at all. First, why would it be on the road? The best explanation we have thought of so far is that somebody is systematically exposing human witnesses to certain scenes, carefully designed to convey certain images.

HYNEK: That's something important.

VALLÉE: The naive UFO investigators say that's evidence that we are being visited. According to them, these space beings come here in their craft and they have to repair their equipment! They come here, and we can see them breaking branches and picking up stones. That's evidence, according to these people, that we're being visited by space explorers, because when we go to the moon we pick up stones too, and we bring them back, and now they do the same thing. But why would they be doing it again, and again, and again? For twenty-five years!

HYNEK: Perhaps they are trying to tell us something symbolically; on the other hand, if they are so smart, why don't they tell it to us directly? Why didn't they land in Boianai? Maybe it isn't transmittable in terms of language. They seem to give you a little bit of an idea and then give you reasons against it or give you something contrary, to change your belief structure.

VALLÉE: The inoculation technique?

HYNEK: The inoculation technique. So maybe this is cosmic inoculation.

HASTINGS: If people believe they are overhearing something inadvertently, they are more likely to believe that information and accept it than if a person is telling them the same thing directly. If you come up on someone on the highway, with the hood of his car up, you're more likely to pick him up than if he is just standing there holding up a thumb. So, one of the aspects of a scientific study really depends upon how you can interpret what you think you study. The hypothesis very subtly influences the method. The only way you can avoid that is by not becoming too attached to how you see what you see. You accept it within brackets, perhaps.

VALLÉE: One point to bring up is that the appearance of UFOs has been verified in all countries, and that different cultures react differently. We have discussed New Guinea, but I think we should also say something about UFOs in China. People traveling in Red China are asked by Chinese people whether UFOs are just a Chinese phenomenon! I know a man from Hong Kong who has traveled widely in Red China, and he confirms that they have waves of UFO sightings there. I have it also from another man, an Australian businessman, that even in very, very small communes in the south of China where they don't even know what's going on in Peking, people have come up to him and said, "What are those things we see in the sky? Do you see them in other parts of the world too? Or is it just us?" The same thing in the Soviet Union. I spent several weeks there, spoke to people, and found they had as much of an awareness of UFOs as we do in this country. It isn't an awareness through Keyhoe's or Menzel's books; it's an awareness of things that are personally heard, from people they know who have seen the objects, although there is no center in Russia where people are instructed to report this. They would not call the newspapers or the police or anything like that, but they are aware of it. The different cultures reacted differently to this. The French started scientific investigations very early; in 1961, I was involved in the creation of a private group for the investigation of UFOs. It was not seeking support or funding but was simply a loose network of scientists spending their personal time and money on investigation. That can be done on a French scale because France is a small country. With ten people who know what they are doing, you can get all the important information; you can do a lot of good investigative work, covering the French territory. In the U.S., ten people couldn't even cover California! There are different cultural reactions to something that is essentially the same phenomenon.

HYNEK: This hits me more than anything else—this Chinese peasant asking the traveler, "Do you see these things in your country too?" That really is strong.

Hynek Goes to New Guinea

VALLÉE: Do you want to talk about anything from your travels in New Guinea, when you investigated the Father Gill Case? What was it like to be there?

HYNEK: I did find Annie Laurie, who had been among the witnesses. I met with her through an interpreter, and I found four or five other witnesses from the Father Gill case—Ananias, Dulcie.... They showed me how the UFO approached and where they looked and so forth.

HASTINGS: What made you decide to make a special trip?

HYNEK: Two reasons. First, it is really one of the great "classics" in UFO history. And second, I had had a long correspondence with Donald Menzel, who thought he could explain the sightings by natural causes, and I wanted to check his theories.

HASTINGS: Where were you coming from?

HYNEK: I took a little trip from Port Moresby to Rabaraba and then from there this little inboard boat took me over to Boianai, and there wasn't a port or anything there, just beach. There are no white men there anymore—just Blacks. I had Father Cruttwell with me. It was a funny thing as we got there, all these natives on the shoreline looking at me and I at them. Finally they got one outrigger out and that didn't work, so about fifteen minutes later they got a larger outrigger out, but that couldn't make shore, so finally they carried me in bodily. The only thing was that the natives were quite suspicious; they thought I was an official police officer of some sort! At first they didn't want to talk; they thought they were going to get in trouble by it. Cruttwell finally got them to talk to me.

One of the things that Menzel suggests with the Father Gill case was that, well, after all, he was the "Great White Father" and these natives would do anything he said. When I told that to Father Gill, he just laughed! He explained that the reason he was sent there was that the mission had fallen into a bit of trouble—there was rebellion on the part of the natives and the last thing the natives were willing to do was to believe anything that the White Father said to them! They were in open rebellion. "How could Menzel say that?" he asked.

VALLÉE: Who is their national authority? Local authority?

HYNEK: Well, now it's undoubtedly the black priest, but . . .

VALLÉE: They have a native priest?

HYNEK: Yes, Anglican. He's a product of the mission.

VALLÉE: Do they have something like local sorcery or something equivalent?

HYNEK: Yes, they most certainly do, because I learned about some special stones that Father Cruttwell told me were evil stones; they are used in instances where in another culture they would get maybe voodoo dolls and stick pins in. Anyhow, these stones have some magic property, and you either pray to them or think about a person or touch a person with it.

VALLÉE: After they saw the thing and waved at the men on the deck of that big craft, since Father Gill didn't have any explanation, did they go to a local sorcerer or wizard of some kind to get an interpretation?

HYNEK: I don't know. At that point they may very well have, because if there's one thing the Anglican Church has been trying to stamp out it is the medicine man's power.

When Father Gill described it, I said, "How is it possible that you were able to go in to dinner when that kind of thing was still going on?" He said, "It was no stranger to me than if the Air Force had been demonstrating a new type of device. I thought it was some new device of the Americans." It was interesting; odd, but he had the same reaction I had when I photographed my UFO.

Here I was, in a commercial aircraft at 33,000 feet, suddenly seeing a strange object that I could not immediately identify. Oddly enough, I didn't think of it as a "UFO." I had my camera in my bag, under the forward seat, and I thought, "Here is a good chance to practice in case I ever see a UFO!" And so I tore out the camera, focused, and shot two pictures, and timed myself. Even though it appeared strange, I never thought of it at the moment as anything but manmade—perhaps a new test device of some sort. I didn't pause to ask myself, "What's it doing up here at 33,000 feet?" The fact that it was seen in bright daylight prevented much of any feeling of mystery. Naturally, I wondered what it was, but that it might be a truly alien object just didn't occur to me at the time. It wasn't until later that I said to myself, "Well, gee, I can't explain this thing!" Father Gill had that same reaction. It was an interesting thing, maybe some new American device!

The Progress of Research

HASTINGS: Given what you said about UFO observations worldwide, and interest in the subject in Western Europe, China, and Russia, what can you say about the progress of research in other countries?

VALLÉE: I think few people realize the amount of work that is going on in other countries, especially in Europe. In France, there are now several groups that are doing much more than mere investigations. Most people in the U.S. seem to think that working on UFOs consists of taking a tape recorder, driving to the country, and talking to witnesses. Well, it doesn't stop there; it only begins there! After you've got a report from a witness, what do you do with it? That's only the beginning, that's not the end of the study! In France, there are several groups that have been conducting analyses for many years. They have gone as far as establishing networks of photographic stations, and magnetic detecting stations. Those have had a surprising rate of success in detecting peculiar phenomena. People have installed those detectors in their houses: If the detector goes off they go out and sometimes actually observe an object. Those are people who have been told what to do if they do see a UFO—trained to recognize the weather pattern, note the temperature, remember the direction the object came from, the exact time, the day, and they centralize all that.

HASTINGS: I've heard of those detectors that work on magnetic principles.

VALLÉE: Yes. Most of those are like a modified compass; it has been observed very often that the needle of the compass deviates when a UFO is nearby. So they build a device that has a stop on either side of North, such as ten degrees on either side, and as soon as the needle touches that, it goes off, it rings. Some of them are automatic in the sense that they keep a record of the time when this went off, so that even if nobody is around you can still see how often the device has detected something. There may be natural things that happen to make it go off—a truck going by, for example—but when you go out and see something in the sky, it's another matter!

In France, for the last two years, the Gendarmerie (which is the local police force, somewhat similar to the National Guard) has officially been given the task of thoroughly investigating UFO cases, *in particular the landings.* Now the Gendarmerie is a very unimaginative, very bureaucratic, but very thorough type of police force; if a landing has been reported, there might be fifty Gendarmes spending the night at the site on the chance it may happen again or some clue may be found.

As far as we know, there are studies of UFOs going on in the Soviet Union.[2] In Spain there are several private groups that have done research, and there is a series of publications from them.

---

[2] See Appendix D.

HYNEK: The work that is being done about the problem in those countries is pointing up the fact that the U.S. isn't doing anything. Other countries are ahead of us in researching this phenomenon. This is something people in the U.S. don't realize. That's why we started the Center for UFO Studies.

VALLÉE: In March of 1974, the French Minister of Defense, Mr. Robert Galley, gave a twenty-minute statement on UFOs while he was interviewed on the French radio. He said that it was essential to keep a completely open mind:[3] the Gendarmes had been studying UFOs for over two years,[4] and their evidence was consistent with that from military radars and the French Air Force. He said there was just too much consistency in those reports to brush aside the question of the existence of UFOs. On that basis, he essentially said, we have to conclude the phenomenon is real and scientists should be investigating it. It is the first time that a cabinet member of one of the major powers has made a positive statement on the subject.

HYNEK: When I was in Canberra, I talked to the chief of the Royal Australian Air Force Intelligence, and his chief lieutenant or aide got me aside and said that he was flying a Sea Fury one night, and noticed that there was this bright illuminated thing off his left wing. "I wasn't going to call in because I didn't want them to think I was crazy or something, and I was watching it for a while. Then another one showed up *off my right wing,* and I was getting pretty damn nervous. I decided to call the tower." In calling in the most unexcited, calm way he could, he merely inquired whether they had any other traffic in his area and they came back, "Yes, two others. In fact, you are the middle blip," and he looked

---

[3] "There has been an extremely impressive increase in the number of visual sightings of luminous phenomena, sometimes spherical, sometimes ovoid, travelling at extraordinarily high speeds ... my own profound belief is that it is necessary to adopt an extremely open-minded attitude towards these phenomena. Man has made progress because he has sought to explain the inexplicable ..." Robert Galley, French Defense Minister, France-Inter broadcast, February 21, 1974—interviewed by J. C. Bourret.

[4] "What can we do, as Gendarmes, when we face this problem (i.e., the UFO). Let us attempt to propose an answer here. The *Gendarmerie Nationale* is in an appropriate position to bring valuable help in the search for the truth in this field: because of its decentralized organization that reaches into every rural town and village; because of its knowledge of the country and especially of the population; because of the intellectual integrity and unquestioned honesty that characterizes its personnel; and also because of its capability for intervention at the site. "How can this search be conducted? By being attentive to reports, by a sympathetic attitude towards bona fide researchers who investigate on behalf of serious, recognized organizations; by questioning witnesses carefully, and without preconceived ideas; by describing carefully the state of landing traces. By doing all this we can contribute to the solution of one of the greatest mysteries of all time." *Gendarmerie Nationale,* No. 87, 1971 [translated by Vallée].

at me, grinned, and said, "You know, I could have kissed those radar guys!" It's things like that—personal stories and anecdotes that build up one's knowledge of this phenomenon.

Radar Confirmations

HASTINGS: Another question that does come up is the combination of radar and visual sightings. People are not yet aware that so many have been made in all countries.

VALLÉE: One thing has always seemed interesting to me; namely that congressmen were told that there were no unidentified radar cases in the air force file.

HYNEK: We should bring that out.

VALLÉE: This was not under oath.

HYNEK: No, but the next thing to it. I mean, after all, it was a Congressional hearing and here he was, the head of Project Blue Book, saying there were no radar cases that were unexplained. I almost felt like getting up and saying, "You lie!" but this was Congress and I didn't.

VALLÉE: There are at least half a dozen good, officially unexplained radar cases in this country alone. There was also a case in Italy in November 1973, involving a private plane that was going to land on the main runway at Torino Airport and was practically told by the tower, "Don't land, there's a UFO on the runway." There was an object hovering, a large egg-shaped glowing object that was directly above the runway. This was seen by several airline pilots and by the crew of an airplane that was going to Rome, and it was picked up on radar by the military airport where the commander got out and saw it visually, confirmed to the press that he had seen it, and that it had been on radar. It followed that airplane for about forty miles.

HYNEK: Here we're dealing with the real world. These are real men of authority, intelligence men in Australia, military officers in Italy, the French defense minister. The real people who're doing these investigations, like the Gendarmerie in France, they've got their feet on the ground.

VALLÉE: Well, the Gendarmes have no imagination whatsoever, they just do whatever they are told to do—only the facts, and they get them no matter what. They just go in there with their equipment and do what they are told to do. And if they are told to spend the night outside watching for UFOs, they'll do that and nothing else. They ask no questions, just do it. They'll fill out the form in triplicate, no matter what happened, and that's exactly the kind of an investigation you want.

Psychic, Natural, or Technological?

HASTINGS: Okay, I have a point right here that was triggered by this conversation. The public seems to put the UFOs, ESP, Uri Geller, all that together on the occult shelf at the local bookstore. People I know whose tastes run that way have linked all these phenomena together without any kind of objective discrimination.

VALLÉE: I think that it is important for us to accept that these are unexplained phenomena, and certainly there is a strong case here for psychic aspects of UFOs. But just saying that UFOs belong to the realm of psychic phenomena doesn't explain them. Uri Geller doesn't explain UFOs, and UFOs don't explain Geller. It's a cop-out, and it should be labeled as such! It is simply a way of saving you the trouble of going there and investigating the cases one by one. If it's a purely psychic phenomenon, why did it land in Kansas? Why did it destroy a wheat field? Why did it leave this substance on a farm in Nebraska? The work that needs to be done is for us to gather that kind of information and investigate it and correlate it and understand how it works, whatever it is!

HASTINGS: There are two senses in which people think UFOs are psychic—one view is that they are created by mental or unconscious projections; in other words, they are not real in the nuts and bolts sense. A second way is to assume that they are parapsychological, in which case they may very well be real, but they are conforming to a different set of laws. In the first case, it would be unlikely that they would leave residues; and in the second case, they could very well leave residues, although the purpose of leaving them would still not be clear.

VALLÉE: Suppose you were in late 17th-century France and attending a gathering of French aristocrats. It was very fashionable in those days to experiment with static electricity. Everybody had his little machine with a glass plate and you would wind it up, the ladies would touch it and there would be a spark, the ladies would scream and everybody would be amazed. Well, that was just amusement, but scientists could well have said this was all magic: it didn't conform to any of the laws they knew. Here was something that would make sparks fly through the air. Sometimes the machine would attract a piece of paper, but if you used the fur of a cat, then it would repel paper. It would produce the opposite effect for no apparent reason at all!

HASTINGS: When were the beginnings of what we could really call electricity?

HYNEK: Shortly after Benjamin Franklin.

VALLÉE: Yes, in 1750 or so, people were already experimenting.

HASTINGS: I would dispute that they would call that magic.

HYNEK: I think you would call these electrical experiments interesting natural phenomena; unexplained, but natural.

HASTINGS: They didn't even call mesmerizing magic, and that was far weirder.

VALLÉE: Mesmerizing has been called magic. There were several books written against Mesmer, saying his experiments were all the work of the devil.

HASTINGS: Well, I don't know—it's reproducible on demand.

VALLÉE: So is the devil!

HYNEK: One of the main differences is that UFOs seem to be under some sort of intelligent control, and natural phenomena are not under intelligent control.

HASTINGS: Give me an example of intelligent control. Other than the occupants.

HYNEK: Well, they apparently exhibit what would be called theatrics, as in the Boianai sightings. Or a light seems to be examining something, moving around an object as though it were examining it—also the trajectories are not random, discs will hover and then take off rapidly. Upon discovery, in the cases of humanoids, the tendency is just the opposite to contact. The reaction seems to be to run away. These cases do not bespeak a natural phenomenon.

3
The Scientists at Work

Unknown Objects in Space and in the Air, and What Science Is Doing about Them

"I believe UFOs belong to someone else and they are from some other civilization."

—ASTRONAUT GENE CERNAN, January 4, 1973, at a Los Angeles press conference

"I believe UFOs, under intelligent control, have visited our planet for thousands of years."

—ASTRONAUT GORDON COOPER, July 1, 1973, at Cape Canaveral, Florida

"Odds are that UFOs exist."

—ASTRONAUT JOHN YOUNG, November 28, 1973, at a Seattle, Washington, speaking engagement

(ALL QUOTES AS REPORTED IN THE PRESS)

Table of UFO Sightings by Astronauts

(From a list compiled by Mr. G. Fawcett.)

February 20, 1962——John Glenn, piloting his Mercury capsule, saw three objects follow him and then overtake him at varying speeds.

May 24, 1962——Mercury VII: Scott Carpenter reported photographing firefly-like objects with a hand camera and that he had what looked like a good shot of a saucer.

May 30, 1962——X15 Pilot Joe Walton photographed five disc-like objects.

July 17, 1962——X15 Pilot Robert White photographed objects about thirty feet away from his craft while about fifty-eight miles up.

May 16, 1963——Mercury IX: Gordon Cooper reported a greenish UFO with a red tail during his fifteenth orbit. He also reported other mysterious sightings over South America and Australia. The object he sighted over Perth, Australia, was caught on screens by ground tracking stations.

October 3, 1963——Mercury VIII: Walter Schirra reported large glowing masses over the Indian Ocean.

March 8, 1964——Voskhod 2: Russian cosmonauts reported an unidentified object just as they entered the earth's atmosphere.

June 3, 1964——Gemini IV: Jim McDivitt reported he photographed several strange objects, including a cylindrical object with arms sticking out and an egg-shaped UFO with some sort of exhaust.

October 12, 1964——Voskhod I: Three Russian cosmonauts reported they were surrounded by a formation of swiftly moving disc-shaped objects.

December 4, 1965——Gemini VIII: Frank Borman and Jim Lovell photographed twin oval-shaped UFOs with glowing undersides.

July 18, 1966——Gemini X: John Young and Mike Collins saw a large, cylindrical object accompanied by two smaller, bright objects, which Young photographed. NASA failed to pick them up on screens.

September 12, 1966——Gemini XI: Richard Gordon and Charles Conrad reported a yellow-orange UFO about six miles from them. It dropped down in front of them and then disappeared when they tried to photograph it.

November 11, 1966——Gemini XII: Jim Lovell and Edwin Aldrin saw four UFOs linked in a row. Both spacemen said the objects were not stars.

December 21, 1968——Apollo VIII: Frank Borman and Jim Lovell reported a "bogie"—an unidentified object—ten miles up.

July 16, 1969——Apollo XI: This was the mission on which a UFO reportedly chased the spacecraft.

November 14, 1969——Apollo XII: Astronauts Pete Conrad, Alan Bean, and Dick Gordon said a UFO accompanied them to within 132,000 miles of the moon, preceding them all the way.

A Jealous Phenomenon

The above series of observations by astronauts was the point of departure of our third session.

HASTINGS: What about various Apollo missions and the UFOs the astronauts saw? During Apollo 11, Neil Armstrong, Edwin Aldrin, and Michael Collins said they observed a UFO.

HYNEK: The astronauts? Some of the NASA movie frames that I examined were most interesting—particularly those taken on the Apollo II flight,[1] one of the few for which NASA has not come up with some sort of explanation. And several astronauts have stated that they definitely saw things in space which they could not identify. Thus—it satisfies the definition of UFO! Unidentified!

Dr. Franklin E. Roach,[2] whose work covered only the earlier sightings by astronauts B.C.—Before Condon[3]—mentions those visual sightings made by astronauts in orbit that in his judgment remained unexplained. In fact, I've always admired how Franklin got himself off the hook by closing his chapter in the Condon Report by saying that those unidentified astronaut sightings remain "a challenge to the analyst." Very neat. He didn't attempt an answer, and as for suggesting that they might really be UFOs—well, that apparently was a no-no.

HASTINGS: Did they say they saw something ahead of them?

HYNEK: Yes. Apparently Neil Armstrong was asleep at the time; I was told that the other two saw something that looked like an open book in the great distance, but when they looked at it through binoculars (they didn't photograph it, in that particular case), it was a cylindrical object.

Another related topic is the fact that people say, "Well, if these things are real, why don't our infrared satellites, why don't our radar networks, why don't our sky surveys, why don't the weather satellites and all that pick these things up?" Well, of course, we don't know whether they do or not. I certainly know that in the satellite tracking mission,[4] we got a number of things that appeared on the films that were never tracked down; they weren't part of the mission! A person who says that the Baker-Nunn cameras never picked up anything is just dead wrong because I know they did. I was in charge of the project! We just didn't bother about it. It would have been too much work to investigate some strange lights; it would have diverted men from the job they were supposed to do; this undoubtedly is true in

---

[1] See, for example, NASA photo #56663402, a single frame from a movie sequence (R. Emenegger, *UFOs: Past, Present, and Future*, Ballantine). It shows an object with some sort of structure in outline.

[2] An American astronomer, member of the Condon Committee, who wrote Chapter 6 of the Report, "Visual Observations Made by Astronauts."

[3] Dr. Edward U. Condon, Project Director of the work, which was reported in "Scientific Study of Unidentified Flying Objects," 1968, Bantam edition.

[4] Dr. Hynek was associate director of the Smithsonian Astrophysical Observatory from 1956 to 1960 and headed the United States satellite optical tracking program.

the case of radar. Radars do pick up all sorts of cockeyed things, but the military figures they're not of interest to a particular mission, so heck with it! I have a wealthy friend in Texas whom I thought I might interest in this research, but his objection was that if the UFOs were real, more people would see them at any one time. He asked me, "Why is it such a localized thing? Why do two or three people here see it and lots of people around the town in a populated area don't see it?" Well, how do we know they don't? Lots of people don't report! I've had a number of cases where people have said that they didn't say anything about it because they fully expected to see it all written up in the papers the next day, and then were surprised to find that there was nothing in the paper about it! But, on the other hand, the UFO is what has been termed a "jealous phenomenon." A Boeing 747 is not a jealous phenomenon, an eclipse isn't jealous, anyone can observe it. But a UFO is a "jealous phenomenon" in that it seems to show itself preferentially in a particular area; a UFO seems to be localized in both space and time.

Why Aren't There Some Pieces?

HASTINGS: One question that comes up is, if they're around, why aren't they leaving some physical residues? Why haven't they dropped a few nuts and bolts?

HYNEK: Ah, that comes up time and again! Why isn't there any hardware left behind? Surely they must crash sometimes, surely they must. . . . Well, I think we should bring these things up. Sure, these are the obvious objections. We don't know the answer to that. Yet, there are a few cases where actual physical remains have been reported—but very few, and these have never been followed through to final proof. Not only would this require very ample funds but, well, what would constitute "final proof"? On the other hand, if you were a bushman in Australia, how many parts from a Boeing 747 would you be picking up? The bushman might see an airplane fly over his territory for many, many years and not once be "fortunate" enough to be witness to an air crash. Think of the thousands of commercial planes flying daily over the U.S., yet years go by without a single crash.

HASTINGS: That is one of the baffling things, because we do know that they cause physical effects: the circles on the ground, broken branches, and so on. Why, if they cause physical effects, don't they leave physical remains at some time or another?

VALLÉE: Accidents have been reported involving UFOs, and material has been picked up and has been analyzed and, apparently, was unusual.

HYNEK: Yes, but always someplace like Ubatuba, Brazil. Why not at least Oshkosh? Or Bangor, Maine?

HASTINGS: Are any of those fairly reliable?

HYNEK: APRO[5] thinks that the magnesium sample from Ubatuba is highly reliable. But that was looked at by the Condon Committee. One of the good things that the Condon Committee did was to analyze that thing, and APRO had to admit that this magnesium wasn't pure; it was highly pure, but it was not unlike magnesium that the Dow Chemical Company produces. There was no greater purity than that of the Dow Chemical. On the other hand, APRO came back and said, "... but the impurities that were in it were quite different from the impurities that you find in the Dow kind. There were trace elements there which were quite unusual." Well, I don't know whether we can build a case on that.

VALLÉE: It does substantiate the story of the witness though. His story is that he saw a disc that flew over the beach and appeared to be in trouble and exploded. And he picked up this sample on the beach in Ubatuba, Brazil, which is close to the standard of purity of magnesium.

HYNEK: And then, time and again, people bring up the Spitzbergen, Norway, rumor. There was supposed to have been a real crash there, and the Norwegian government reportedly investigated it and confiscated everything. High government officials are said to have given credence to it. But to get something on that—I don't know how you're going to do it! Those are things in the past. I think we can say that those things ought to be revived, but we don't have the facilities to persuade the Norwegian government to give us the data.

The Men in Black

VALLÉE: What about the rumors about Men in Black?

HYNEK: The Men in Black, the three Men in Black?

VALLÉE: It has been frequently reported, especially in cases involving landings and photographs, that shortly after the sightings, strange men would appear, sometimes flashing Air Force credentials—or NORAD credentials or police credentials—confiscate the evidence and disappear. However, this cannot simply be explained with the idea that they are from some hush-hush place in the government or elsewhere, because they had knowledge of the sightings before the witness made a report. Also, there have been witnesses who said, "I don't want to say anything because these men showed up and told me that I would be in trouble if I said something."[6] Usually, several men were involved, dressed in dark suits, or at least business suits.

---

[5] The Aerial Phenomena Research Organization, a private group with headquarters in Tucson, Arizona.

[6] For a detailed example of a case of this type, see Chapter 5, "The Night an Occupant Was Shot."

HYNEK: And driving a black car, or an official-looking car.

VALLÉE: Government types. Your standard image of typical FBI investigators. Very businesslike, asking all the right questions—no nonsense. It's very nice as a science-fiction story, but apparently it did happen in several verified instances. There is a case in Damon, Texas, where two deputy sheriffs who were driving saw a UFO near the road and tried to get close to it. They got scared and drove back into town; the UFO chased them. By that time, they were slightly panicked and they drove as fast as they could. One of them had been bitten by a pet alligator earlier in the day and had an open wound on his arm, which was resting on the edge of the car window while he was exposed to the light of the object. The UFO then receded rapidly. When he got into town the wound was healed completely. It had been a pretty bad bite. He went into a local restaurant with his buddy and ordered a drink. A strange man was there, got up, went to their table and said, "You know, if I were you, I wouldn't say anything about that thing you saw on the highway." And he left. They hadn't told a soul about it. That's kind of typical of a whole range of things. These are incomplete patterns. Have you found any indication of that in talking to witnesses or reports from police?

HYNEK: A few times, yes. A few times the witness said, "I was told not to say anything about this," or in two cases a guy said, "Somebody from the Air Force showed up and made me give him the photos." And I have been told by several Air Force officers that gun camera photos of UFOs got confiscated when they got back to base. That sort of thing, but not very often. These things are very hard to document. In twenty years, things get sort of lost, strayed, or stolen.

Who Should Study UFOs?

VALLÉE: Perhaps someone is taking the problem seriously after all, not necessarily the U.S. government. And by the way, if you were called by the president to give him a briefing on the UFO situation, what would you recommend? How high would you put it as a matter of national priority? Is it more important than the energy crisis?

HYNEK: I would say it might help solve the energy crisis.

VALLÉE: How would it do that?

HYNEK: It would help the energy crisis because if we could determine the propulsion method that they have, then we wouldn't need conventional energy sources; that's what I'd tell the president. There might be a discussion on possible technological spinoffs of studying the UFO problem along those lines. Of course, the energy crisis is such a pat thing!

HASTINGS: Okay, take yourself as a scientist. Let's suppose you have the technology of the UFOs, what would you do with it to pursue your scientific interest? Would you do some moon exploration, would you do high altitude observation, would you do weather studies?

HYNEK: We could use that technology to explore the whole solar system. Here on Earth it would have many applications—communication, construction, transportation, crop control, long-range weather forecasting....

VALLÉE: It would have military applications.

HYNEK: It would certainly have military applications—if you could find out, for instance, how they can stop cars, you could have a whole army advancing towards you, and you could stop all their vehicles.

HASTINGS: You know *The Day the Earth Stood Still*—that movie? Okay, let's suppose you're a psychologist or sociologist; what would studying UFOs mean, how would that help you in your study of psychology or sociology?

HYNEK: If we could learn something about their method of communication—for instance, apparently they can implant ideas and messages without going through verbalization, hearing, and reinterpreting. Think how much is lost if I have an idea and I have to put it in words, you hear it in words, and then you have to retranslate it! If I could project the idea directly, what a tremendous help that would be!

VALLÉE: A thought amplifier.

HYNEK: Amplifier, yes, and thought projector—stereo, yet! Now one thing we should ask is to whom does this problem belong? What sort of scientist should work on it? Does it belong to the physicist, to the psychiatrist, or to the anthropologist? Are UFOs an interdisciplinary subject? We don't know to whom the problem belongs. Any institute that gets set up to study it should have not only physical scientists but also social scientists, and it should have medical men, since there are burns, nausea, headaches, all sorts of things reported. That is why we have different kinds of scientists associated with the Center for UFO Studies.

VALLÉE: A lot of investigation work is just a matter of routine police work. We need good detectives more than scientists.

HASTINGS: It is a lot like the other subject areas we've talked about—in astronomy, you have an overlap between physicists and astronomers and I suppose chemists and others....

HYNEK: Oh, yes.

VALLÉE: But what's interesting is that in the normal course of science, when a new subject comes up, like ecology, everybody wants it! The sociologists say,

"It's *our* subject, *we* should be studying it." The biologists say, *"No, it's our subject."*

HYNEK: Yes, because they all want the funds for it.

VALLÉE: Yes, they want the money!

HYNEK: You just get some funds for this UFO subject and everybody will want it.

VALLÉE: With UFOs, so far, it has been exactly the other way around. You have articles published by psychologists saying, "This obviously is explainable by an astronomer." The astronomer says, "All these people are lying, they should be seeing a psychiatrist!" The psychiatrist says, "Just look at those traces on the ground, the physicist should be studying that." And naturally the physicist won't touch it!

HYNEK: That's a good point.

VALLÉE: They just pass the buck—science passes the buck.

HYNEK: Passing the UFO buck. . . .

HASTINGS: But the ironic thing is that if there were bucks available, nobody would be passing them!

Is There a Cover-Up?

HASTINGS: Another question you always get asked is: What about the government, what do they know about this? What are they covering up? Are they covering up?

HYNEK: To me, it is inconceivable that in any government like ours, which spends so much money on intelligence work, somebody wouldn't be interested in this. I approach it from this angle rather than trying to get proof that it is or isn't a coverup. It's inconceivable to me that it wouldn't be of interest to somebody on top, and if it isn't, then somebody is being damn dumb!

VALLÉE: It may be another case of passing the buck. You could extend what I was saying earlier, with the astronomers saying, "Well, the intelligence community should be studying this!" And the intelligence community may be saying, "Well, it's the Air Force that should be studying that"; and the Air Force says, "We can't study it, because the scientists tell us UFOs don't exist." Remember, every time the military gathers a panel of prestigious scientists, as they did in 1953, the scientists say, "Heck, we pay you guys to protect us from the Russians, we don't pay you to go around spreading superstitious stories about little green men!"

HYNEK: That's another case of passing the buck.

VALLÉE: What about the Air Force? My feeling is that those people have been rather sincere in their attitudes. The last thing they want is to be laughed at by a committee of Nobel Prize winners—there is no way that whoever is in charge of a project in the Air Force can seriously investigate something when the most prestigious scientists in the land say the thing doesn't exist!

HYNEK: That's a good point and should be brought out. We still have to say that it really isn't completely the Air Force's fault if the scientists have told them there's nothing to it. In my years of experience with the Air Force, I was impressed with the regard and awe Air Force officers held top scientists, yet the same officers felt superior in practical matters to these men whom they regarded as impractical academicians. But they were ready to accept—even blindly—their word on scientific and nonmilitary matters. The matter was brought to the Scientific Advisory Board, and they never regarded it as a serious matter. You must remember that UFO stories are really so bizarre and unbelievable in everyday terms that it was quite natural for scientists to reject them. They are used to battling crackpots and pseudoscience on an almost daily basis!

VALLÉE: You were close to the 1953 panel, and both of us were close to the Condon investigation in 1967. What would have happened if in either of those two cases, say Condon in one case, or Robertson[7] in the other, had said, "This is the most important subject that could be studied by scientists—there is much knowledge to be gained from a study of the subject." What would have happened?

HYNEK: There would have been a large group set up immediately. Instead of the orders from the top to play it down, to debunk, we would have been a hell of a lot further today than we are. It would have been followed up, had Alvarez and Robertson at that time simply said, "Look, there's something of potential value here." I talked to Robertson's sister; she is a physicist at Montana State University, and I spoke there. When I said there was never any scientific work done on UFOs, she asked, "What about the Robertson panel?" Well, I didn't disabuse her, I said, "Well, yes, but they only looked at the cases for about five days and it's pretty hard to do a scientific job in five days!"

VALLÉE: People often believe that Robertson and the other scientists were deliberately hired to debunk UFOs. But there is evidence that they were sincere. When we spoke to Dr. Alvarez a couple of years ago, he was still saying there

---

[7] In 1953, the U.S. government assembled a prestigious panel for a five-day examination of the Air Force data on UFOs. It was chaired by a physicist, Dr. Robertson, and issued the statement that the phenomenon deserved no scientific study and should be "debunked."

was nothing to UFOs. And Condon on his deathbed was still saying there was nothing to UFOs. All these scientists are consistent in every way in their negative position.

HYNEK: Well, I was saying there was nothing to it myself when I started! Look, for several years I was saying there was nothing to it. I was in the midst of these reports off and on for months and months, yet I remained what at the time seemed to be reasonably skeptical—much too much so as I see things now in retrospect. I desperately tried to find rational explanations for even the most bizarre reports. Well, then, if these guys look at it for just five days, why should we expect them to be anything but skeptical?

I thought the whole thing was a fad, a craze—and would pass from the scene as fads invariably do. Back in 1948, when I first started, I would have taken just about any bet that by 1952 the whole matter would be forgotten. It was the persistence of the phenomenon, not only in the United States, but over the world, that finally grabbed my attention.

HASTINGS: What has science done about UFOs then? Is this the extent of it?

HYNEK: What science has done about it? Well, if we consider the Condon Report, science hasn't done much about it.

HASTINGS: I think you go a little far in attacking the Condon panel. Maybe you have a grudge against it. Your grudge is perfectly justified, but it strikes me that there ought to be a way of presenting it in less personal terms.

HYNEK: What you mean is that I was mad because I wasn't on it or something like that?

HASTINGS: Yes.

HYNEK: The reason I wasn't on the committee is that the Air Force wanted a "fresh start"—excluding anyone who had previous acquaintance with the subject. The same is true for Jacques.

VALLÉE: There is a point that should be brought out about the Robertson and Condon panels and government cover-up. The people who say today that the government knows everything about UFOs have been saying this for the last twenty years. Now, if the government knew everything about UFOs in 1947, then you would expect certain things to have happened since then.

HYNEK: Such as what?

VALLÉE: Well, would we be spending so many millions of dollars on jet fighters and rockets?

HASTINGS: Research directions would have changed.

HYNEK: Well, that's a very good point. That is the best argument against cover-up that there is, because there would have been some spinoff from that cover-up. We're at a sort of impasse because it's inconceivable to me that any government that has such a strong military intelligence arm would not be doing something about it. And yet, on the other hand, if they did know all about it, then there would have been some technological spinoff from it, and we would have had much better ways of getting to the moon, among other things.

HASTINGS: Well, of course, that assumes that we could understand the technology of the UFOs.

HYNEK: Perhaps they are trying to do something about it but they don't understand the nature of "UFO technology" yet?

HASTINGS: What are the methods that could be used to understand their technology? What tools do you have at your disposal?

Three Avenues to a Solution

HYNEK: The three topics that come to mind quickly are the computer, photographic analysis, and magnetic detection.

VALLÉE: What do you think of the usefulness of a computer system that would give you information on sightings?

HYNEK: First of all, the computer gives you instant access. You don't have to go thumbing through all sorts of papers. This was the trouble with the Air Force files! If you wanted to find out what landings occurred in South Dakota for instance, you had to go through every one of the papers from the year 1947 to the present, because they were arranged only in chronological order! I tried many times to get the Air Force to put them in a machine-readable form. Even to use an elementary cross-indexing, which would have been tremendously helpful, but was never done until Jacques did it in 1967.

Everybody and his brother speaks glowingly of the use of the computer, but everybody has a different idea. NICAP[8] has pursued a project called "Access." APRO started in the direction of using a computer program, but I haven't heard how far they've gotten. In any case, the Center does have a computerized data bank going.

---

[8] The National Investigations Committee on Aerial Phenomena, based near Washington, DC.

VALLÉE: One thing we should do first maybe is to demystify the computer approach by pointing out that computerizing everything is one way to stop research by introducing a complete sclerosis. To devise a standard code and put everything on the computer in a coded form and do nothing else might be disastrous. The reason for this is that, if you're in the manufacturing industry, and you have twenty-five different kinds of nails, you can encode them from 1 to 25 so you can have a complete inventory of nails, right? But now if you're trying to encode or produce statistics in a domain where you literally don't know what it is you're studying, it's a different ball game! What you want to get *out* of the computer *is not what you put into the computer in the first place!* Then it's a dangerous fallacy to say, "We're simply going to dump all this in coded form into the computer and do statistics and we'll get the answers out of the computer." It's literally garbage in and garbage out.

HASTINGS: Well, what should be done along those lines? What is productive?

VALLÉE: What is productive is not a large database approach. The *first* thing to do is *to publish all the case information in plain English.* Those who centralize that information should feel the responsibility to publish it as soon as possible and to make it available to other people. It's definitely a fallacy that the group of us here, or any other group of people anywhere, is going to find *the* answer to the UFO problem, computer or no computer! An answer is only going to come when large numbers of people are given the facts and can think about the facts and do their own research, bring their own input. Then we'll have some semblance of progress. But first, the information should be published by the groups that are gathering it in any form. What the witness said, not the coded form. With as much information as possible.

HYNEK: Here's where we do have something positive to suggest to the private civilian organizations. We might suggest a positive program that they can use. I hope also maybe we can warn them about the Mystique of the Computer. A good example of what can be done is the UFOCAT system that Dave Saunders has created. He has computerized reports from various sources. Of course, he does it without prejudging their validity. He is especially interested in the dates and geographical distribution of the reports.

VALLÉE: People have been brainwashed into thinking that in order to do anything with the computer you need millions of dollars. They think you need your own computer, and you need to hire programmers, and you need to write programs, and you need millions of dollars and many years to code the information, and so on, and that is *just not true.* What you need is a small-scale but very well organized effort to use the most modern techniques of content analysis. You want to use those techniques from the actual words of the witnesses and not on some code. In the code, you only lose information. In a search of the landing catalog Allen and I did recently with my terminal, we processed a request on "close encounters involving dwarfs." Well, the computer retrieved (among other things) one case where the witness described "a dwarf with furry ears"! What are the chances that some programmer creating a code would have anticipated this? Can you imagine a programmer saying, "Oh, well, we are going to need a code OX3H for the dwarfs with furry ears!" This is the kind of information that is lost in any code, where you don't keep the actual words of the witnesses. Yet, who knows? It may be important.

A Computer Scenario

HYNEK: Writing a scenario for the processing of UFO reports, let us suppose that the private UFO groups do agree to cooperate with us in a unified scientific program. Suddenly we are flooded with reports. NICAP gives us ten thousand reports, and APRO gives us another umpteen thousand, and so forth; now how much do you estimate it would cost to put that into the form you suggest? Certainly, not with all these fancy cabalistic codes. We don't want to translate everything to that darn code. I agree that we want that in English. Now you have more experience than anybody in this country with coding the workings of the English-language computers. Tell us how that would be done now. Or suggest some approach.

VALLÉE: If you want to do it right, and *if you restrict yourself initially to close-encounter cases*, you would need a budget of $70,000 to $90,000 a year, with about $30,000 of that spent on computer time and terminals. You need a staff of maybe two people with good language backgrounds, assigned to summarizing the cases in English as we get them, and then quickly getting the summaries printed and published in batches. I've used that technique in two instances, working on a shoestring budget. The first time, I used it for a catalog of landings that I published in *Passport to Magonia* as an appendix. I did this by hand, initially, taking all the cases and summarizing them. I ended up with 923 cases of landings. Having this, I duplicated it and then started to send copies to about half a dozen people around the world who were doing similar things, mainly people who were interested in landings and had been cataloging them. There are a handful of people like that around the world; I know who they are, and I correspond with them. So I sent them my list and they checked it against their own lists in the different countries. Then they added some information; they would say, "Case number X turned out to be a hoax." So I would remove it. They added some new cases too. Then I would update the entire list, print it out, and send it around again.

HYNEK: Did you send any of those to NICAP or APRO or MUFON?[9]

---

[9] MUFON is the Mutual UFO Network.

VALLÉE: No, because they didn't have anybody doing catalog work. There is no catalog of the APRO file or of the NICAP file that I know of; all their data are in filing cabinets. This catalog went to people in England, Spain, France, and Italy who are doing catalog work on a continuous basis. After two or three rounds of that, I ended up with a catalog that I published. Later, it was translated into different languages and field investigators in the different countries took the cases in their area and went to reinvestigate them. I am now collecting the data of the third generation information; after twenty years people have gone back to find a witness, saying "We have this report which was published in this catalog. Can you tell us more about it? This happened twenty years ago when you were a kid. Now you have a family and a responsible job, what can you tell us about it? Was it a prank or was it serious?" Thus, we have a new extension of the information base. I think that our model has been tested, and now we can do the same thing on a larger basis.

In 1971, I did a similar study with Ballester-Olmos in Spain, building a catalog of cases of close encounters with UFOs in the Iberian Peninsula. This time we used the computer more extensively. He sent me his notes, and I sent him my own lists. Practically all of his reports were unpublished outside Spain, and two-thirds of them were even unpublished in Spain. This was a lot of information that had been collected privately. By that time, I had developed a technique for content analysis on the computer, so we put everything on the machine and produced chronological listings that I could send back to Spain every week. I would also produce a listing sorted by region in Spain so that he could tear a certain one out and send it to his field investigators. Every few weeks, his field investigators in the different regions had an up-to-date list of the cases they were working on. About two or three times a week, I would get a list of updates from him, and I now have about fifteen hundred pages of raw material pertaining to those hundred cases of landings. At the end of this process, we had these landings investigated and reduced to summary form, where all the information was put back into the computer in English again. The computer could manipulate the text very fast and made it possible to publish that information and print it without having to do the retyping and sorting by hand. And I think we can now do that on a larger basis.

What I'm talking about is using the computer in support of the intellectual ability of the human brain. It's not that the computer is going to simply print out a statistical matrix or a bunch of numbers or correlations of statistics or something like that. Take a case, for example, where an object has just landed on a beach and was seen by two fishermen. You want to know how often does this happen? Or someone said, "I saw a UFO that was shaped like a triangle and gave off green light and there was something that looked like a dog that was seen near it." Well, you want to know in how many cases something similar happened. Where the witness may use a word or expression that is very unusual, you want to know if there is anything similar to that in the file. So it's very different from a statistical, mathematical routine. It's more a case of supporting an intelligence operation.

HYNEK: How much does a typical search cost?

VALLÉE: Given modern computers, the cost is by no means negligible. For example, just to keep in storage all the information about close-encounter cases costs over $35 a month. Take a typical session where I would—

1. Extract all the cases from Chile and justify the lines of the text.
2. Extract all the Spanish cases, save them, append the Chilean cases and justify again to have a nice printout.
3. Return to all the Spanish cases and generate an index by region.

A task such as this takes me roughly fifteen minutes at the terminal and the cost is about $8. Take this as an order of magnitude. In a few years, with better equipment available, the cost will be only a fraction of this.

HYNEK: Still, this is very small compared to other means of obtaining the information. Think of how many hours it takes to compile and print an index manually.

VALLÉE: There are several misleading proposals saying we need two million dollars to start a computer database, and three years from now you'll get some UFO statistics out. This is simply unrealistic. We have to warn people that this is not the way to go.

HASTINGS: Well, what you're saying also about not really needing this gigantic funding situation is very much in agreement with what some of the big daddies have said about cancer research. These people at many of the institutes that have been doing significant cancer research are saying, "What we *don't want* is a large central authoritarian cancer research institute," which is what we have now. They want little pockets of people working on their own, not being directed by one central authority. Not one single money-controlling authority.

HYNEK: Certainly. I think we can almost say, if the history of science has any bearing on it, that when the solution to the cancer problem comes, it is not going to come by direct attack; it's going to come by some little run around left end by somebody we least expect.

HASTINGS: But you do need to have the material, the reports. They have to be available to enough people who want to spend that intellectual energy on the problem. And that seems to be the device of a central storage computer.

VALLÉE: I think many people are becoming aware of the dangers in the "big databases." They are concerned about social security databases and the police databases and so on, and I think the history of this field is littered with the remains of the big multimillion dollar projects for databases which have never led anywhere. Many companies literally spilled millions of dollars into creating completely useless files, piling up data into the computer without having any idea of what it was going to be used for, or how they were ever going to retrieve it. I think the fallacy there is that we think we can store information in the computer when all we can store in the computer is *data*. *Data turns into information only when someone asks a question about it!* But until you know the questions, you can't define a code! So, it is true that if you're in the manufacturing business you can define everything you're ever going to want to know about nails and screws and bolts and nuts, but if you're doing research, by definition you're analyzing something the nature of which is unknown; then you cannot anticipate that, and the idea of this big database project is a fallacy, and people should know that it is.

HYNEK: That needs to be brought out certainly, because I know that several UFO organizations have gotten me aside and asked if I would be a consultant to them on computer business and so forth, but then I saw what they were up to; they wanted everything in numerical codes, a six- or eight-digit number that had everything in it. That's just nonsense. This brings us back to what the Center for UFO Studies is and what it is not. The Center should be more of a clinic than a center . . . in the same sense of what the Mayo Clinic is to the medical world. It should not try to put other organizations out of business in any sense of the word; on the contrary, it should help them. We should try to reinforce them whenever possible, but we should also contribute to disseminating the information to the public.

The Analysis of Photographs

VALLÉE: After computers, you said the second important topic was photographic analysis. How do you evaluate pictures? What do you do when somebody reports taking a photograph of a UFO? Why are pictures so important?

HASTINGS: Most people feel that pictures are concrete evidence that there was really something there. There was a physical object taking up space and having mass, and the picture really proves it. "A picture is worth a thousand words . . ." that seems to be the feeling.

VALLÉE: Yes, but on the other hand no picture can *absolutely* prove it. You can fake anything!

HYNEK: It is proof only if you take it yourself.

VALLÉE: But then this is only convincing to yourself, so the others are no better off.

HYNEK: A picture also can serve as a good illustration of what others have described. For instance, the Calgary disc. It is a beautiful illustration of the typical descriptions, "Silvery hamburger sandwich," "Two saucers that kids ride down the hills on, one on top of the other." We have just verbal descriptions ... and suddenly we come across a photograph that looks like that. While I can't say that this is genuine, it certainly does illustrate what people have described.

VALLÉE: Yes, but then you're open to the argument that maybe somebody read the description and built a model. Take the old Adamski[10] saucer with the three spheres under it; well, people have been making models like that, flying them and photographing them and selling them! So sometimes they're put forth as proof that Adamski was telling the truth, but it's just somebody who has read his book and made one.

HASTINGS: Let me ask something about photos here. If you were able somehow to obtain a strip of film out of the same camera, same lens and same time sequence and same lighting condition, and within a few frames you had not only a photograph of a saucer but also of an airplane, approximately at the same distance. Then you analyze the filmstrip itself. Would the quality of the image of the airplane on the film be any different than the quality of the image of the saucer?

HYNEK: It would be the same; if the distances and size were really the same.

HASTINGS: Has that kind of analysis been done?

HYNEK: Yes, to this extent it has been done. People like Adamski who have tried to fob off a model that is a few feet in front of a camera can easily be detected.

HASTINGS: By the focal length?

HYNEK: Not the focal length, but the atmospheric haze. If you had a real flying saucer at the same distance as the plane, both images would have the same optical history. The clearest giveaway on the fake picture is that it is always sharp and distinct, everything stands out. To have a good fake, you should have that model a mile or so away; it wouldn't stand out sharp and clear.

HASTINGS: Suppose you accepted as genuine all those photos with which you don't find anything wrong. Your intuition doesn't tell you there is something wrong with them, and they seem to meet most of the criteria. Now suppose you make the jump and say, "Okay, let's accept these as genuine." What kind of data would you be able to get from that kind of analysis?

VALLÉE: They confirm what people say concerning shapes and structures. You don't really learn anything more.

---

[10] George Adamski, a California man, claimed to have met with people from Venus. See Chapter 6.

One thing that I have noted is that although there are only a few pictures that show distinct objects with a good degree of credibility, there are a large number of pictures where the subject was close by at first and then moved away, and *the witness caught it when it moved away.* In such cases, a visual description or a sketch is much richer than a photograph. So all you have is a blob of light. People tend to reject that. But you could get other types of data from the photograph, even from a poor picture. People say, "Well, we can't use that because it's just a blob of light, it doesn't show any details!" Yet the whole science of astronomical photography is based on the analysis of blobs of light! There is an incredible amount of technology that has been devised exactly for that.

A thing that I'd like to see done would be for someone to collect maybe twenty photographs showing just lights in the sky that we have good reason to believe were UFOs and then analyze them the way we analyze galaxies, by isophotometry. There is a piece of equipment that will give you lines of equal light intensity. From that you can do very detailed analyses.

HYNEK: This is both costly and time-consuming, but it is something we hope can be done by the Center for UFO Studies.

VALLÉE: You could also do photometric analysis where you would get the luminosity profile of one of those things and from that you could see whether it has sharp edges, whether there are different levels in the light; is the center darker, or are the edges darker? From this you can infer a lot about the source of energy that produces the light. And all this can be done. It would be much more interesting than a closeup picture of a dark flying saucer with three windows, because the flying saucer with three windows doesn't give you any clue as to the type of energy. But when it's in a lighted state, you could find out a lot. This is the kind of technology that is available and yet no one has done that.

HYNEK: "All this can be done." How easy it sounds! I agree, it should be done, but not even the Condon Committee, with a half-million dollars, did it. Research requires support—both financial and intellectual. It amounts to just that.

HASTINGS: You know, I think at some point someone should publish information like that; things that have yet to be done, but need to be done and could be done.

VALLÉE: Yes. Here on one side we have the pictures and on the other side we have the equipment, and it's only a matter of taking the pictures and feeding them through the equipment. You don't need to go out into the field, you don't need two million dollars, you don't need anything. But you do need good documentation on the conditions when the pictures were taken, and you have to centralize all that carefully. Of course it's not as simple as I make it sound, because in astronomical photography, along with those blobs of light, you do have calibration.[11]

HYNEK: You know, a sketch sometimes appears more honest than a photograph. If you have faith in the person, if you're dealing with his testimony, then why not deal with his sketch? This is as honest as he could be! Of course, being honest is not the same thing as being accurate. An *authentic* photograph would clearly be best, but.... It's odd, a sketch can be honest but not accurate, a photograph accurate but by no means necessarily honest!

Magnetic Effects

HASTINGS: Now the third thing you mentioned, Allen, was the magnetic effects. I'm wondering whether it makes sense to discuss correlations of sightings with conditions in the magnetosphere of the earth; who has studied the magnetic condition ... with respect to sightings at that place on earth?

VALLÉE: The closest thing to that, to my knowledge, is a study by Claude Poher, based on data from magnetic detectors.[12] The French have a network, as this country does, for atmospheric research in general, and also for detecting nuclear explosions; it is based on extremely sensitive detectors that monitor the three components of the local magnetic field. Poher has been able to correlate apparently random variations in the magnetic fields with periods of intense UFO sightings in France (see Figure 15). Unfortunately, the likelihood of having a sighting very close to a magnetic station is very low because those stations are usually in isolated areas, in order to be free of electronic "noise." The main one is in the center of a forest in France where there is no house within twenty-five miles. However, knowing the distances of the sightings to the station and the maximum peak-to-peak amplitude of the disturbances of the magnetic field on that day, you can try to draw a graph of how that correlation varies. If you extrapolate that curve, you get a very, very high value for the magnetic field at the site of a UFO, which is consistent with everything else. If we are faced with a technology, it involves very large values for the energy and the electromagnetic field, all generated within an object of fairly small dimensions.

---

[11] An astronomical photograph is "calibrated" by exposing a small area of the film to a series of light sources of known intensities for a precise amount of time prior to taking an actual picture.

[12] Claude Poher was director of scientific experiments for the French National Center for Space Studies in Toulouse.

82 THE EDGE OF REALITY

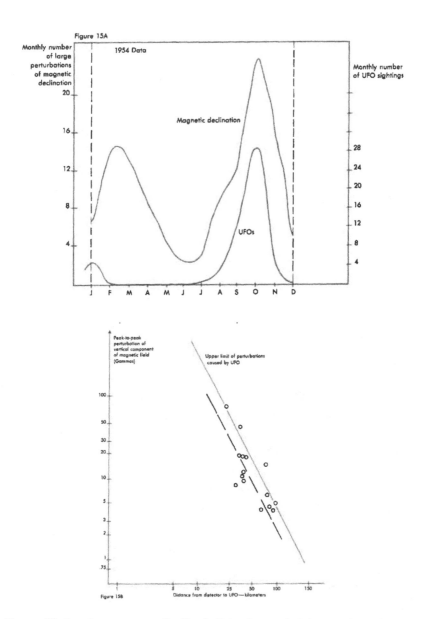

Figure 15 French space expert Dr. Claude Poher has studied the correlation between the number of UFO reports during the 1954 wave of sightings and the number of large perturbations of the Earth's magnetic field (15A) and of the variation of the amplitude of magnetic perturbations with the distance of a UFO (Figure 15B). Based on magnetic field measurements at Chanston-la-Forêt (France). From C. Poher, "Correlations entre enregistrements geomagnetiques et UFOs" (July 1973, private communication).

4
"It Would Be Too Frightening to Many People"

The Hill Case, Hypnosis, and Pascagoula

In October 1973, two fishermen in Pascagoula, Mississippi, observed a low-flying object and its occupants, and subsequently one of them described his "abduction" into the craft. Thirteen years earlier, two other people, Betty and Barney Hill, had had a similar experience in New England. In both cases, the witnesses were put under hypnosis, and it was found that most of the information about the sightings had been repressed by their conscious minds. In the Pascagoula case, both witnesses became so distressed when asked to relive their experience that the attempt had to be terminated. The recurrence of such incidents raises a number of fundamental questions—what happens to a person who has a close experience with these objects? What can we say about its reality? How good an analysis tool can hypnosis be?

Encounters with Occupants

We had an opportunity to explore these questions in a discussion organized by Chicago radio station WIND on March 31, 1974, during which we were able to interview Mrs. Betty Hill and to exchange ideas with a Chicago-based expert on hypnosis, Dr. Larry Garrett. The discussion was moderated by Mr. Ed Schwartz, of station WIND, who first asked us what we thought of the Betty Hill case in relation to other UFO sightings.[1]

VALLÉE: What is especially striking to me in this case is the description of the encounter with the "occupants." My own interest is centered on cases of close

---

[1] Transcribed from recorded WIND broadcast, Chicago, March 31, 1974.

encounters and the type of reality that the witness seems to enter. Betty and Barney Hill described their car being stopped on the road. The craft apparently was on the road directly ahead of them, blocking the road, and they saw five occupants coming toward them. At that point, they lost control of their own volition. In other words, they could no longer decide to run away, to escape. They described being taken aboard the craft and being given a medical examination. Now, of course, we have to ask whether this is something that took place in our normal reality, or how we can reconcile this account with our everyday concepts. And this is the question that none of us has answered yet, I think.

SCHWARTZ: In other words, this incident has never been explained away by anybody as being not true.

VALLÉE: That is correct. The striking thing to me is that there were several distinct parts to the experience and they were reported differently. There is a part that they reported immediately. They remembered seeing a light; Barney had looked at it with binoculars and thought he had even seen a window on the craft. And this was communicated to the Air Force. However, there is a second part of the experience they reported only to their immediate relatives and friends; that was the loss of time involved in the sighting . . . the two hours where they literally didn't know consciously what had happened. And finally there is a part of the experience that their unconscious mind had not reported to their conscious mind, if you see what I mean. That only came out under hypnosis. *That is the part that concerned the occupants of the craft.* By the way, when going through the Air Force files with Dr. Hynek later, we found that the Air Force had a radar report from the same night. So there had been radar contact with an object in that area.

SCHWARTZ: [to Garrett] Why don't you take this a step further and tell us how regression is done and what it means, if you will.

GARRETT: I've probably done about four hundred or five hundred regressions, if not more; regressions are a way of taking subjects, hypnotizing them, and having them recall past events. You might call it a form of hyperamnesia. What you do by recalling these former events is this: If you have a subject who responds to a deep enough state of hypnosis, he will actually relive what he has experienced in the past. I didn't get to hear these tapes, unfortunately, but I would imagine that on the tape that Dr. Benjamin Simon made, the regressions were in the same tone of voice as when this craft supposedly landed in front of the car; possibly Dr. Vallée could go a little bit further on that. But I would imagine this is what happened. In regression, you can actually take a person way back in time. I've even had some people who have been back to the age of three months and have been able to describe the furnishings in their bedroom at that age when they were, in fact, say thirty to forty years old, with the mother sitting by the side saying, "Yes, that's true, and we moved away from there shortly thereafter and there would be no way for him to know this." Regressions are phenomenal.

SCHWARTZ: Are they always accurate?

GARRETT: No, they're not. A lot of times people use their imagination. A lot of times people fabricate things, from either wishful thinking, fantasies, dreams, things such as this. If you've got a somnambulistic subject, one in a deeper state of hypnosis, you can get a good true regression and you can almost rest assured that you are going to have an accurate regression. You're not going to have too much made up in the case of something like this. But you can see what Dr. Simon did: He had Betty Hill at one time and Barney at another, so they could not say, "Well, that's not true," or "It's like this." There was no note comparing. So that shows that the regression was accurate. Anybody who is into hypnosis and does any type of regression would find out that many times people have such a vivid imagination that they will sit there and make up all kinds of things just to please the hypnotist. But in a regression like this, there's very little doubt but that it was a good regression. I would say there would be no doubt whatsoever, because he interviewed them separately.

SCHWARTZ: What happens is that the subconscious mind acts almost like a videotape recorder, then?

GARRETT: Well, the subconscious mind does retain all meaningful materials from the time we're born, and any meaningless material supposedly would be discarded, but I have worked with subjects where I have been able to bring out even such trivial material as looking at a street sign. Many authorities say all meaningless material is cast aside, but I have found many subjects who have recalled the temperature at the time, the weather, whether it was raining ... like a videotape that has recorded what has gone on from the time you were born to the present day. Through hypnosis (this is one of the beneficial uses of it) you can actually go back to the time and pull out any program you want to see....

The Uses of Hypnosis

HYNEK: Could I ask Dr. Garrett how he uses hypnosis in his clinical work?

GARRETT: You mean in what category?

HYNEK: Yes.

GARRETT: In all of them, Dr. Hynek. I teach it at two of the colleges here in Chicago. I have two offices in the Chicago area. I work with about twenty-three physicians who send me patients for nontherapeutical and therapeutical applications, such as treating allergies, smoking, weight loss, etc. As far as regressions, I have done quite a bit of research on my own and with other people. Right now I'm doing a lot of research with an ophthalmologist in the area of hypnosis and vision.

HYNEK: Well, I certainly want to talk with you more when I get back to Chicago.

SCHWARTZ: An interesting point, Dr. Hynek, is that hypnosis might apply to other sightings, people you have interviewed, for example, who were unable to remember minute details. Don't you think?

HYNEK: I should certainly say so, and one place where I'd like to see that applied would be in the recent case in Mississippi, in Pascagoula, where it would be very interesting to bring out what the subconscious mind of Charlie Hickson and Calvin Parker have recorded.

SCHWARTZ: Has that close encounter ever been fully explained, or is that still under investigation?

HYNEK: It's never been totally explained, and I'm hoping that we can mount a further investigation of it.

SCHWARTZ: Now, Mrs. Hill, I'd like to place your location when this sighting took place on September 19, 1961. Where were you and your husband when this incident occurred?

BETTY HILL: We had left Montreal in the afternoon, and we were driving down through the White Mountains of New Hampshire on Route 3. This cuts through the mountains, and the sightings took place over an area of approximately thirty miles.

SCHWARTZ: Had you ever had an experience with UFOs prior to this?

BETTY HILL: No. In fact, we knew almost nothing about them.

SCHWARTZ: Your husband had no knowledge either, then?

BETTY HILL: No, he had heard the term but had discounted the whole idea. He was very closed-minded.

SCHWARTZ: Were you a skeptic yourself, or were you open-minded?

BETTY HILL: I had no opinion.

SCHWARTZ: At what time of the day did this occur?

BETTY HILL: We estimated the UFO started to move in close to us right around 3 A.M.

SCHWARTZ: Was this a reasonably lonely stretch of highway where this occurred; were you alone at the time, no other cars around?

BETTY HILL: It was strictly a resort area . . . the summer season had ended and all the hotels and motels were closed . . . we had started on Tuesday night around midnight, and we had driven without seeing a car for most of the time.

SCHWARTZ: And what happened is that a light appeared before your automobile and began to grow larger. Is that it?

BETTY HILL: First it started following along beside us for several miles and then left the top of the mountain and came out over the highway and stopped in front of us.

SCHWARTZ: Can you describe it for me? How big did it appear?

BETTY HILL: It was a large craft.

SCHWARTZ: Was it as big as a house, for example?

BETTY HILL: I would say it was fairly close to sixty-five feet across.

SCHWARTZ: Now what happened? At some point after this it approached you, your car stopped, is that what happened?

BETTY HILL: Well, when it stopped in mid-air, my husband Barney was driving, and he stopped the car to get a good look at this object.

SCHWARTZ: What did you think it was? Did you have any idea?

BETTY HILL: I thought it was a UFO.

SCHWARTZ: What did Barney think?

BETTY HILL: Well, he didn't believe in UFOs. He was determined he was going to try to identify this craft.

"They Are Going to Catch Us!"

SCHWARTZ: What did you both do then?

BETTY HILL: At this point, he left the car, taking the binoculars with him to get a good look at it, and when this happened the craft shifted out over the field, and there was a red light on the side that seemed to move out from the main structure. The craft started to move toward us, and we could see a double row of windows and this human-type figure standing behind the windows.

SCHWARTZ: Now in the classic description we've heard of UFOs, was this a saucer-shaped device, a cigar-shaped device—what did it appear to look like?

BETTY HILL: It was a disc shape.

SCHWARTZ: Now after Barney left the car with the binoculars, you were still in the car at the time?

BETTY HILL: Yes, I remained in the car at that time.

SCHWARTZ: What happened next?

BETTY HILL: Well, Barney was looking up at those humanoid figures; he identified one of them as the "leader," and he had the feeling that the leader was telling him to "stay right there, don't move, no harm is going to come to you." And Barney broke away and ran back to the car, yelling, "They're going to catch us, we'd better get out of here!" He got in the car and the motor was still running, and he shifted into gear and drove down the highway for a while, and at one point there was a series of beeping sounds; the car vibrated and then the car would go on, and later I said to Barney, "Do you believe in flying saucers?" The car vibrated again and then we returned home. It was later, in going over that, that we realized that there was a time lapse of about two hours, during which we had no recollection of what we did.

SCHWARTZ: What happened then? You went home, and you realized there was a two-hour time distortion, and your husband was nervous and upset about this?

BETTY HILL: Yes. He later developed an ulcer.

SCHWARTZ: Over what period of time did this occur now? How much later?

BETTY HILL: This happened in '61, and we went to see Dr. Simon for the first time in December of '62. Barney had been under treatment for several months.

SCHWARTZ: Then it was a period of two years before the psychiatrist first had a chance to interview you both, right?

BETTY HILL: Yes.

SCHWARTZ: And had you discussed it much during those two years? Had it still been preying on your mind?

BETTY HILL: Only relatives and a few close friends knew about it.

SCHWARTZ: Until you went to the doctor.

BETTY HILL: Right.

SCHWARTZ: Now you explained exactly what you told us to Dr. Simon, right?

BETTY HILL: Barney was going to a psychiatrist locally to see if there was any emotional basis for his ulcer . . . and he was worried. The doctor referred Barney to another psychiatrist and also recommended that I should go along too. It was through Dr. Stevens that we arrived at Dr. Simon.

SCHWARTZ: Did you find either Dr. Stevens or Dr. Simon skeptical of this description initially?

BETTY HILL: Well, I don't . . . Dr. Simon wasn't, but I think basically he probably had almost no interest in UFOs. I think his attitude was, as a psychiatrist, to try to come up with some kind of psychiatric answer.

SCHWARTZ: But what happened was that hypnosis was used, and this information came out during hypnosis.

BETTY HILL: Right.

SCHWARTZ: Were you quite surprised to find what had happened under hypnosis?

BETTY HILL: Actually, when Dr. Simon played the tapes back to us, it was quite painful.

SCHWARTZ: What was it on the tape that made it painful? You discovered that you—and I use the word "alleged," being a journalist here—that you were taken aboard this "alleged" craft and your bodies were physically examined. . . . Is this the crux of the story here?

BETTY HILL: Yes.

SCHWARTZ: What supposedly happened inside this craft now?

BETTY HILL: After the craft came down, the car motor died . . . the men separated, came up one on each side of the car, they took us out and they took us into the craft. They took us on board. They took Barney into one room, apparently for examination. Later they told us we were going to forget the whole experience, took us back to the car and sent us home.

SCHWARTZ: How would you describe these creatures? Or these humanoids . . . what did they look like?

BETTY HILL: Well, they were short. I'd say the leader was the tallest one. He was about five feet. Their features were different from ours in that the eyes were very, very large, and they had almost no nose. The skin had a greyish tone to it. They were all dressed alike.

SCHWARTZ: What ever happened to the tapes of your regressions, yours and your husband's?

BETTY: Dr. Simon has a copy, John Fuller has one, I do, and there's a copy in the Library of Congress in Washington.

SCHWARTZ: Is this material something that you have released to the press? Have these ever been heard on radio or television?

BETTY: No.

HYNEK: Betty this is Allen, and Jacques is on the line also. [Greetings exchanged.] I've heard the tapes and so has Dr. Vallée, and I can assure you they are really of historical value.

SCHWARTZ: Do you believe the public should ever hear them, Dr. Hynek?

HYNEK: I think under special circumstances, yes, but not just for general curiosity.

SCHWARTZ: Mrs. Hill, what is your opinion here? Do you think the general public should have access to them?

BETTY: I think they would be too upsetting and too frightening to many people.

HYNEK: Yes, I think they would out-exorcise *The Exorcist*!

The Pattern

SCHWARTZ: Dr. Hynek, would you say this has been the closest encounter you have ever dealt with?

HYNEK: Yes, it very definitely is.

SCHWARTZ: What is your opinion here? She's described the craft. Is that just a classic description you've heard before?

HYNEK: Yes.

VALLÉE: The description is quite classic. There have probably been ten cases of close-encounter situations of that type in the last four or five years, wouldn't you say, Dr. Hynek? I don't mean cases that would involve necessarily that kind of dialogue with the occupants. One thing that was very striking to me when Betty described it and when I read it in John Fuller's book[2] is the nature of the dialogue. It is both very logical and at the same time it has a certain component of absurdity ... it's absurd or dreamlike in parts, but at the same time it is consistent. It is definitely not the kind of thing that would come out in an ordinary dream.

SCHWARTZ: An important point to make here is that this occurrence has never been explained away as anything common and ordinary.

HYNEK: No, it certainly hasn't, and as a little personal note here, I think one of the greatest experiences in my life was the time that Dr. Simon put Betty and Barney under hypnosis in my presence. They were both kind enough to allow this, and for an hour and a half I talked with Betty and Barney while they were sitting in front of me under hypnosis. I learned a great deal, not only about the incident, but about hypnosis itself, and the things that can be done.

SCHWARTZ: Let me bring Larry Garrett, who is our hypnotist, back into the conversation. Larry, let me ask you something. Could an incident like this be recalled over and over and over again under hypnosis and would it be consistent?

GARRETT: Yes, it could. Every time the person is rehypnotized, and every time the person is regressed again, the same exact incident would come out again. Of course, luckily they were both deep-trance types. This would mean that over and over we could regress Mrs. Hill, and she would come up with the same exact thing

---

[2] *The Interrupted Journey* (Dell, 1967)

that happened in 1961, and I have never found anybody yet who has flawed. I regressed one particular individual twelve times, and he never once flawed on any of his examples of what had happened in the past.

VALLÉE: How many people are deep-trance subjects like that?

GARRETT: Well, almost any book you find and any authority will say 20 percent of the population. But the only thing I disagree on is that in my own work I have found that the more I work with a subject, the deeper I can get him, so through conditioning, you could take this 20 percent and raise it higher. We could take the most analytical of people and condition them to become a deep-trance type within a certain number of hypnotic sessions.

VALLÉE: To what extent is regression material acceptable in a court of law? If not, then why not?

GARRETT: It's not ... there's the thing. Hypnosis is just not accepted as it should be ... yet the benefits that are derived from hypnosis would take years and years to discuss. I agree with both you gentlemen and Mrs. Hill on the point of not allowing the general public to hear the tapes, because they would probably fear it or laugh at it. It seems that anything people do not understand, they are either afraid of or they scoff at it. Hypnosis seems to be one of the things that people do not understand. They won't accept it, the courts of law won't accept it. I've been working with different police departments trying to get them to accept it.

SCHWARTZ: That's a very interesting point. I know that Dr. Hynek and Dr. Vallée have other questions. I have other questions for them too. Dr. Hynek, as far as the scientific community is concerned, you and Dr. Vallée have credentials that are impeccable ... how about your colleagues in this area? How have they responded? Have they tended to discount this incident?

HYNEK: More and more of them are getting open-minded. We now have a Center for UFO Studies that is comprised of faculty members from several universities, and I would say the scientific community is slowly coming around to accepting these matters in the same way that Dr. Garrett said hypnosis is being more and more accepted.

SCHWARTZ: Do you think there's any discredit in the fact that two years elapsed between the incident and its first coming out under hypnosis?

VALLÉE: I think it's quite normal that there would be this delay. We still see people coming to us now with incidents that occurred many, many years ago. Sometimes they have told absolutely no one about the occurrence because it was something that was too personal or too deeply disturbing for them. I think it's quite natural.

SCHWARTZ: Dr. Vallée, have your investigations and studies continued although you've moved to California?

VALLÉE: Yes, they have. I've been studying the evolution of this problem in this and other countries very closely. One thing I'd like to mention, since you were talking earlier about the skepticism of the United States government on this subject, is that one month ago the French Minister of Defense released a statement (see page 59) that the reality of the phenomenon was undeniable on the basis of military radar observations, pilot observations, and the investigations of the local police.

SCHWARTZ: How about the Soviet Union?

VALLÉE: We know of work going on there, but it has not yet been publicly discussed. It's the private work of a few scientists, but we know this work is going on.[3]

SCHWARTZ: Dr. Hynek, do you think the Soviet government might ever join with the United States, or at least allow their scientists and you and your colleagues to get together and compare information?

HYNEK: It's quite conceivable, in good time.

SCHWARTZ: There's a good possibility they might have some very valuable information.

VALLÉE: I think every country has information to share with every other country on this subject.

HYNEK: I would like to see a group in the United Nations—perhaps their Committee on Outer Space would be appropriate (!)—act as a mechanism for the exchange of data on the UFO problem between scientists of various countries.

SCHWARTZ: I want to go back to the incident that occurred in Pascagoula, Mississippi, in the fall of last year. Now, between 1961 when Betty and Barney had their experience and this incident occurred, has there been anything similar, Dr. Hynek, or are these the two closest together?

HYNEK: To the best of my recollection, not in the United States, but I believe, and Dr. Vallée can probably jog my memory, I think there were a few incidents in the South American countries.

VALLÉE: Yes, and there were at least two cases in France. But every case, of course, is somewhat different. One very strong thing here is to be able to compare the features of observations in different cases, and I'm anxious to hear what Betty thinks of the Pascagoula sighting.

---

[3] See Appendix D.

Betty Hill Comments on the Pascagoula Case

BETTY: That experience sounds reasonable to me. I think there are many similarities, also many differences, too.

VALLÉE: What were some of the differences?

BETTY: Well, these humanoids had claw-like hands and the men were floated into the craft. I was dragged in.

SCHWARTZ: You found out under hypnosis that you were dragged in, right?

BETTY: Right.

SCHWARTZ: When you were dragged in, were you placed in some type of state where you were not resisting after that?

BETTY: After I was inside the craft, I just had the feeling that I had better cooperate and go ahead with whatever they wanted so I could get out of there and go back home.

SCHWARTZ: Was there any language spoken at any time, or was all the communication done through mental means, Betty?

BETTY: My communication was with what I called the leader. He made sounds, and I understood him. When they communicated among themselves, it was a very weird sound.

SCHWARTZ: To summarize . . . we've got an incident here that occurred on the 19th of September 1961, that's obviously still very vivid to Betty Hill and still holds a great deal of interest for people who follow this subject. It may never be explained at all, but you're still going to push ahead, and your new project is going to see what it can find out about all these things, is that correct?

HYNEK: We're getting more and more scientists to take a solid look. There have been twenty-five years of nonsense and now, finally, as so often happens in other areas, a more serious attitude is being taken toward it. I think our new Center for UFO Studies will gain further support.

VALLÉE: May I ask one more question of Betty? I would like to know if you have seen any UFOs since the incident?

BETTY: I have not. But I would like to tell you that last night here in Portsmouth, there was a fairly close sighting of a very large UFO.

HYNEK: That is interesting. I think you know about this; we have a UFO Central hotline that has been set up for the Center to get reports like this directly. They contact me wherever I happen to be in the country when an important sighting takes place. Hardly a night goes by that we don't get one or two reports. It's quite significant.

SCHWARTZ: We should explain here, there currently is a study underway under Dr. Hynek's guidance. The hotline they use has a toll-free 800 number, and law enforcement agencies all over the United States have been given this number. And you say you get calls almost every night. How fast do you check these once they come in?

HYNEK: If it's really a hot[4] case, we can get it checked within a matter of hours ... we have investigators in various parts of the country who have agreed to cooperate and get on to it almost immediately, so it's really quite an interesting system.

The Problem of Contact

Using this review of the Hill case as a springboard, we turned our California discussion to the problem of contact from a physical and psychological point of view.

HASTINGS: When you take away the cases like the Betty Hill sighting or the Pascagoula case, where the person has been paralyzed by fear or otherwise rendered inactive, are there cases where the witnesses were actively involved with the contact?

HYNEK: The Hill case wasn't exactly passive. Barney struggled quite a bit. In listening to the tapes, I was shaken by the terror in his voice as he described his fight to resist the abduction.

VALLÉE: There are cases in Latin America where the witness reportedly started fighting with the occupants. The different cultural reactions to the "occupants" in different countries are interesting to me. In the U.S., the people tend to be frozen with fear, or run away from them. In France, they tend to have intense curiosity, or try to speak with them. In Latin America, the cases have been more physical, more involved. Brazilian researcher Olavo Fontes has confirmed to me one case in 1954 when a man drew a knife and hit the "occupant"; the knife rebounded as if it had hit a metallic surface. Then he picked up the "occupant," who was very light, maybe fifteen pounds, but the thing just hit him, and he [the man] went rolling away as if he had been hit by a truck—so mass and inertia somehow were out of range.

There have been cases in this country where "occupants" have been fired at,[5] and in one case, the Hopkinsville incident, the "occupants" seemed to absorb the energy from the bullets and start glowing brighter.

HYNEK: I think we are frustrated by cases such as the Pascagoula case because they are so damnably unreal, but I am asked about it all the time. People have really been struck by the experience of these two men.

---

[4] That is, judging from past experience, we have good reason to believe the case is potentially truly baffling.

[5] See the details of such an incident in Chapter 5, "The Night an Occupant Was Shot."

VALLÉE: In Pascagoula, we have to ask was the experience real, and if so, in what kind of framework? Aren't we really at the "edge of reality" now?

HYNEK: I think it was a real experience—to these two men at least.

VALLÉE: But perhaps a real physiological occurrence without being a true physical occurrence? As in an erotic dream?

HYNEK: You mean, like a wet dream? Well, in hypnosis someone can suggest you've got a blister on your arm, and you get a blister on your arm.

HASTINGS: They could see this thing—maybe they haven't even talked about some of the events.

HYNEK: Well, I was there. I was an active participant in that case.

VALLÉE: What do you think of the *Rolling Stone*[6] article about Pascagoula?

HYNEK: First of all, those men didn't talk that way. Hickson doesn't have that exaggerated drawl, that extreme Southern accent. I can't even imitate it.

HASTINGS: Where did the *Rolling Stone* get that idea? Obviously, they wanted to take a Southern town apart.

HYNEK: Yes, they sure as hell wanted to take it apart.

HASTINGS: What about their information content in terms of reporting the experience?

HYNEK: Very poor. People ask me, what did they say? What's inside? What was Calvin Parker doing? Well, *they were in a state of shock*, and it was more like a religious experience, as if they had seen the Virgin Mary. They forgot the boring details—can't describe what she was wearing that day, belt buckles or rings on her fingers or what....

HASTINGS: You ought to state the facts as clearly as possible, give the information from your own investigation.

VALLÉE: But one question I hear many people ask is, did it really happen? The Betty and Barney Hill case made more sense. The public could believe that they had been abducted. In Pascagoula, it could have been an out-of-the-body experience—it's like some of the experiences of Bob Monroe.[7] Did it really happen?

HYNEK: Those men think it really happened. That's all I can say.

---

[6] "When the UFOs Fell on Dixie," by Joe Eszterhas, *Rolling Stone*, issue #152 (January 17, 1974).

[7] See *Journeys Out of the Body* by Robert Monroe (Doubleday, 1971).

Pascagoula and the Nature of Reality

HASTINGS: Well, Dr. Hynek, if I were a journalist, I would say, "I want to know whether it really happened or not!" Allen, you're supposed to be the expert, and you should arrive at some conclusion now. You went there; you talked to them; you saw them. Now given your expertise, do you think it really happened to those guys?

HYNEK: I saw no evidence of a physical craft anyplace around, so I can't say whether it really happened. *The men are not lying. I'm quite convinced of that.* I know the men were in a state of shock. Charlie Hickson passed a lie detector test.

HASTINGS: That's not going to satisfy everybody, as you know! What can you say that will protect you and at the same time give people release from the tension they have in asking that question? Can you say, for example, in a court of law that the evidence would be accepted as having proved the abduction?

HYNEK: I wonder whether one could say that. According to the rules of evidence, I think this would be accepted in a court of law. At any rate, the prosecutor could not obtain a conviction that these men were hoaxing. I could say that.

VALLÉE: What were they doing for the whole time of the incident? The sighting lasted a certain amount of time. We haven't established this yet. We need to know what has been happening and what condition the men are in now. We need to follow this case for months, even years. We need a professional medical hypnotist, too. The man who did the hypnosis was an engineer.

HASTINGS: It didn't strike me that he was doing a very sophisticated job.

HYNEK: He was doing pretty well. He could bring them right to the experience. He said, "You are in such and such a shipyard and you see two men sitting on the pier fishing." He was trying to get them to describe it as though they were seeing it in a movie, but even that was not enough. I can remember the tension in the room.

VALLÉE: So we have these two people fishing on the pier. Did the hypnotist reconstruct what happened, minute by minute?

HYNEK: No, it was impossible. The witness became distraught. It seems that he just didn't want to relive that frightening experience, even under hypnosis. But hypnosis can bring out things not in the conscious mind—and in this case, further hypnosis might fill out the same sort of time gap that was present in the Hill case, but apparently much shorter.

VALLÉE: At what time did it all begin?

HYNEK: Nine o'clock, as I understand it. Charlie Hickson snags his line, he's cussing because he lost his bait, and he leans back to his tackle box to get more bait. As he leans back and looks over his shoulder, he sees this blue light approaching very quickly; he doesn't have any idea how fast, but rather quickly there it is in front of him—this apparition of a sort of football, oblong-shaped object, which does not land but hovers about eight to ten feet above over the land (they're fishing off the pier). The thing hovers over this old, broken-down, auto-wrecking yard and even if it had landed, it would have been very difficult to find any physical trace of this because it's just rubble—rock and gravel, essentially—so there wouldn't have been any mud where you could have seen imprints. Three creatures, I believe, appear, but two of them—

VALLÉE: How did they get down?

HYNEK: The two just floated down, he said....

VALLÉE: Was there a door that opened?

HYNEK: A door opened—some sort of aperture.

VALLÉE: Was the door facing them?

HYNEK: He wasn't sure.

VALLÉE: And they came out one by one?

HYNEK: He didn't say that—we can check—and they floated down; it's significant, they didn't crawl down and they didn't jump down, they floated down.

VALLÉE: How far away was it?

HYNEK: I think he said it was about twenty yards or something, quite close. And then two of these creatures came down and each one of them took Charlie by an arm, and he said he didn't feel any sensation from that point on.

VALLÉE: Could he see what was happening to the other guy?

HYNEK: No, he was so scared, he didn't know what was happening.

VALLÉE: Because, you know, that makes one creature on the left and one creature on the right and one creature ...

HYNEK: One creature left for Calvin? But only two descended, Charlie said.

VALLÉE: Unless one of them was grabbing both arms.

HYNEK: But Calvin passed out—that is their story at any rate, and no record of whether Calvin was ever in the craft or not—but this would come out in hypnosis.

VALLÉE: I'd much rather have the hypnosis done by somebody who has not been associated with UFOs before. The good thing about Dr. Simon in the Hill case was that he's very experienced with people in shock. You don't want to select a hypnotist on the basis of his past interest or association with UFOs and psychic investigations. That's not enough, you know, as a criterion.

HASTINGS: Most of them are not experienced in this kind of investigation.

VALLÉE: Another thing, there might be parapsychological transfer during hypnosis. Many people can put them under hypnosis, but we don't know what goes on between the two. The hypnotist might plant things in their mind. I am rather skeptical of the hypnosis Puharich has done with Uri Geller in this respect. People are more receptive to psychic influences when they are under hypnosis, and unless they are dealing with someone who is absolutely professional, who doesn't have a vested interest in the subject himself but is looking at it purely technically, you can get anything.

The Dangers of Hypnosis

VALLÉE: Arthur, what can happen under hypnosis? What are the dangers? There are lots of people learning hypnosis these days, just so they can become instant UFO investigators or instant psychics. They go around hypnotizing housewives and regressing them to their first birthday party or even their "past lives"! Shouldn't people be cautioned against that?

HASTINGS: The most obvious thing that happens under hypnosis is that the person is extremely open to any subtle, unconscious, nonverbal, as well as verbal suggestions of the hypnotist, and they are extremely compliant. If you ask them to go to a past life, and they don't *have* a past life, they will invent one for you! If you suggest that they saw a UFO, they would have seen a UFO.

The other aspect is, the deeper the hypnosis, the more it is moving away from the normal psychological state and the more subconscious and unconscious material is available to the person. So, it is quite possible that without the hypnotist realizing it, material might come up that will unsettle somebody.

VALLÉE: Can you regress someone to such a frightening experience that he might have a heart attack and die?

HASTINGS: I would think it would be possible, but fortunately no cases like that exist—that may be because people don't regress subjects to a frightening experience. The standard way of doing it, of course, by professionals working with things that are frightening, is to get permission of the subject himself, or herself, to do that experience. In effect, they say "Are you willing to go to that incident at this point? Are you willing to remember it at this point?" In terms of trauma, just because you're in hypnosis doesn't mean that it's any less frightening, and, in fact, under hypnosis you might be somewhat compelled to go into it, whereas your defenses would be up in a normal state.

HYNEK: Well, there was an inner struggle, and you could see that in the case of Charlie Hickson—there was a real conflict going on between compliance of the person under hypnosis and things he didn't want to do.

HASTINGS: That's probable. For example, a psychologist I know was working with a man in therapy and under hypnosis. The patient gave a long discussion of how his parents wanted him to do one thing and he wanted to do something else, and that was the source of the problem. The psychologist said, "Well, are you going to remember this when you wake up?" and he said, "No." [The psychologist] said, "Okay, you won't remember it when you wake up until a later point." And so even at that point, he had to retain no conscious memory of that material—it was too traumatic. Emmett Miller, an MD who uses hypnosis to deal with emotional problems, phobias, and behavior patterns, has people in a state in which they reproduce their experiences emotionally, but he has them separate themselves from their emotions so they can see them simply, as in a movie, and that sometimes relieves the trauma. It seems to me that's awfully tricky.

HYNEK: The Betty and Barney Hill case always comes up. I can relate it in terms of how Pascagoula differs from Betty and Barney Hill. In the experience I had when they were hypnotized for an hour and a half, a remarkable thing was the incident in which Betty was sitting to Barney's left, and Barney said to Dr. Simon, "Something is funny. I know that Betty is sitting here [to his right, where she would have been since he was driving], but why is her voice coming from the other side?" So he rearranged their seats, and Barney was happy. Also, one thing that impressed me: When Dr. Simon got through, he went over to Betty and Barney and said, "Now, Betty and Barney, you will wake up, you will feel normal, you will feel happy and you will not remember anything we have been talking about." And he did that and they woke up. Barney didn't like that. He said, "Dr. Simon, I wish you would have let us remember what we talked about because I'd like to ask Dr. Hynek some questions." So he put them under quickly—he's so practiced that they went under immediately—and we went through the same rigmarole except he said, *"You will remember,"* and they did. That was, to me, quite revealing about hypnotism; that you can manipulate the mind that way to complete forgetfulness and suddenly, just by a simple word, you can have them remember.

HASTINGS: Did you say people are using hypnosis for regular sightings? Regressing people and letting them reconstruct it?

VALLÉE: Yes.

HASTINGS: Are they using hypnosis for many sightings? Is it providing information?

VALLÉE: There are quite a few people training themselves. I personally know two or three who are training themselves to use hypnosis, so that if they come across any case where the witness has a time loss, they will be able to regress them.

HASTINGS: *I would certainly advise anybody who is a witness not to be hypnotized*, or anybody who is hypnotizing not to hypnotize anyone unless they get a signed release, and that usually puts people off.

VALLÉE: Why isn't hypnosis used commonly in a court of law? Have you got anything to add to what Dr. Garrett said about that?

HASTINGS: One reason is that not everybody can be hypnotized or at least deeply enough to be useful. Second, it would have to be voluntary. Third, hypnosis is simply not accepted in law. It's been done in a few cases, but it's not generally accepted.

VALLÉE: Suppose I see a car accident on the highway. It happens very fast and later the judge says, "Well, did this car come from the left or from the right, and what was the blue car doing at that point?" I can't remember. "Well, Your Honor, I just don't remember that." Now, in this case I would be willing—I was there, I saw everything, I just don't remember. What if I testified under hypnosis?

HASTINGS: It would be accepted, but in most cases when this has been done, and it has been done, it takes a preliminary statement, discussion by psychiatrists, and so on to establish the validity of hypnosis as a way of recalling that kind of material.

VALLÉE: What are the legal requirements for using hypnosis? Because you were saying we want material that could be used in a court of law.

HASTINGS: Yes, your question really would be to what extent can hypnosis be used, and if it's used, to what extent is it acceptable?

VALLÉE: And to what extent is it equated to the truth? If I say, "Well, the blue car came from the left and it hit the truck on the right fender," is it the truth, or is it not the truth? Do you know more than you knew when you started?

HASTINGS: What guarantee do we have, let's say under hypnosis, you are not simply inventing something that you didn't know in your regular state? I think that's the question.

VALLÉE: If hypnosis is getting big as an investigation technique, we have to clarify its dangers and its capabilities. What are the limits of hypnosis? How much can you say? Suppose you regress me to when I was driving my car coming back from picking up Dr. Hynek at the airport. Could you take one instant and get me to describe the whole landscape? Or only what I was looking at, where my attention was focused?

HASTINGS: You could, in all probability, describe the whole landscape, not only the things that are in your concentrated focus of attention, but anything that was coming in.

VALLÉE: And what I was thinking about?

HASTINGS: I think it would take a very skilled hypnotist to be able to do all of that without interrupting.

The Modern View of the Brain

VALLÉE: How is this related to what modern concepts tell us about the brain?

HASTINGS: Some experts in the field of consciousness think that each hemisphere of the brain is suited for particular kinds of response to the world. The left half of the brain generally processes information in a linear way, using language processes. It is concerned with sequential time, and is involved with logical thinking; scientific, analytic thinking; abstraction; conceptualization; and so on. The right half of the brain seems dominant in spatial recognition, music, sensory reception and perception, movement, gestalt perception; in other words, anything that requires pattern recognition. Global consciousness tends to be in the right half of the brain. It's too easy to say that one half tends to be dominant over the other. What's generally true, though, is that one half operates for certain things and the other for other things.

HYNEK: How long has this been known?

HASTINGS: Fifteen years. The research began about '54 or '55, I think. And it came out of research originally with chimpanzees and then with Sperry's work with epileptics at Cal Tech. When he severed the corpus collosum (the connecting tissue between the two halves of the brain), he found that the subjects developed two consciousnesses, rather than one integrated one. For example, by presenting things to only one eye, which corresponds to one hemisphere, he would get a memory; but activating the other half there would be no recognition of whatever it was, because in these subjects the normal connection between the two hemispheres had been cut during the surgical operation.

To give a titillating example, with one lady, he covered the left eye so she was perceiving things with the right half of the brain. Then he showed her an erotic picture which embarrassed her. He asked her what it was and she said she didn't know and she giggled and finally she said, "Well, doctor, you have a very funny machine!" Her body was reacting, she was embarrassed, her face was getting red, but there was no conscious recognition of what it was because her awareness was centered in her left side. For example, you could present the subject with this [holding up a water glass] when their left side is "on" and they could say, "Yes, it's a glass, of course." Now present it to them when their left side is "off," and they wouldn't know what to call it. They would drink from it, but they could not name it. They might say, "Oh, a pencil?" They would simply be guessing, because the meaning of it would not get through to their language side.

VALLÉE: Its function would still get through, since they would drink out of it.

HASTINGS: Yes. Aphasics are a good example of a natural phenomenon. Aphasics have language deficiencies. They can handle things, they can take pictures with a camera, they can write with a pen, but they can't tell you what it is. A neurosurgeon in Montreal, named Penfield, got similar results in his open brain surgery operations. He pioneered finding areas in the brain which had lesions causing epilepsy. So in doing this, he would touch electrodes to different parts of the exposed surface of the brain with a very low current and would find certain things. For example, he would get complete recollection of some incident in childhood. Also, he would find that by putting the electrode in one place, the person would not be able to give the name of something although he would be able to use it functionally.

VALLÉE: But in his experiments, isn't there also some evidence for being able to trigger higher-level abstract concepts? For example, the *feeling* of approaching but not a feeling of what it was that was approaching?

HASTINGS: I don't know that. I wouldn't be at all surprised. He got a very interesting one which I thought was pretty significant. He would show his patients pictures of things and ask them what they were and in a certain position they couldn't say. One person described it this way. Penfield showed him a picture of a butterfly and his answer was, "I couldn't think of the word *butterfly* so I tried to think of the word *moth*, but I couldn't think of that either!"

VALLÉE: Well, is there any evidence of this in UFO cases? I am thinking of a case in France that has been followed closely. The witness has gone back several years later to see the man who investigated the case; the first sighting happened in '67. The witness went back to see the investigator in '73 to tell him that something new and very important had happened. He had traveled about two hundred miles to the investigator's house, but when he finally reached it, *he couldn't find the words to express what he had wanted to say.* He said, "You know, I can't express what it is

I want to tell you. I have something to tell you; it's important for me to tell it to you, but I can't." And he left, completely puzzled. There was nothing they could do to retrieve it.

HASTINGS: I can think of some connections. Suppose you don't *have* to express. Suppose you don't have to express it *in words*. Suppose you dance it, or suppose you do it with fingerprinting and color, or suppose you do it with gibberish and inarticulate sounds. Suppose you say, "Well, what I saw was, ya, ya, ya, ya." That might express it very well, and you might be able to get correlations between half a dozen people who would listen to you and say, "Yes, if I were doing that, that's how I would say it." Many things that don't come through verbally, in a questionnaire, for example, could come through in another kind of expression.

The Problem with Questionnaires

VALLÉE: My feeling about questionnaires is that they should always have ample space for drawings and should always begin with a white page, a free-flowing essay where people can express what they think happened. The first question would be: What happened? And then in the back, we'd say, "Now please go back and check whether you included the following information: the place, the time, who was with you," and that kind of thing. Instead, the approach to questionnaires is usually to appear scientific! So first you have to give us the duration in seconds and the altitude, the degrees and direction of the compass that it came from; then at the very end they have two miserable little lines for the witness to say what he really feels. It seems to me that that's conditioning the witnesses to remember only certain things, to recall only certain details, and it's doing exactly the opposite of what should be done.

HASTINGS: It would be as ridiculous as an analysis of dreams where you would first ask, "How long was the dream?"

VALLÉE: Yes, and what time of the night was it, what compass direction was your head pointing to when you woke up?

HYNEK: How many glasses of water did you drink? All these descriptions that we have people make, filling out questionnaires and so forth, are really poor. We should do what the airlines do when they're describing lost baggage. They show you a standard set of drawings and ask, "Is it like this? Is it like that?" If you ask a person to describe what sort of bag he had, he will say, "Well, I think it was sort of big," and you'll never get the bag back. But if you simply check the one that's closest to it on the standard set, you will. I'm just wondering if we should have a UFO observer's kit with that sort of standard thing and also a color wheel. You don't ask the witnesses what color it was, you ask them which color corresponds to it the most. That would be something positive.

Ordinary questionnaires have all sorts of fallacies. In fact, in handling the Air Force questionnaire, I found much more information from the narrative account. The Air Force had a blank page in the back for the witnesses to say what they saw in their own words. I would read that first and get the picture, and then I would go to the questionnaire that told me what the time of day was and so on. I like to pick up the total picture. When I'm interrogating the person, I want to build up a mental image as fast as possible and I need the background. So of course, I do ask was it night or day; was it clear, cloudy, and so forth; and were there houses around, were there trees, was it an open field? So I get the picture. Then I say, "Okay, now tell me what happened." I place myself in that picture, and then I can see their description and it gives me a chance to reconstruct the event. If they start giving me a description, and I don't know whether it was night or day when the thing was happening, I'm lost. But I think if we have a recognition set and a color wheel, that will be very functional.

I should mention that questionnaires have gone, and are going through, a real evolution. The mistakes of the past are being corrected. For instance, MUFON now has a greatly improved questionnaire that could well be used by all investigators. In addition to an excellent general questionnaire, which includes ample room for the "what happened" aspect, the investigator is furnished with supplementary sheets, if applicable, to car-stopping cases, physical-contact cases, and animal-reaction cases. There is no use in having a general questionnaire include questions about animal involvement if no animal was involved—but a supplementary sheet asking about what kind of animals, how many, and what the reaction of the animals was is of primary importance when animals were involved.

Hynek's Hypnosis Experiment

VALLÉE: You did an experiment once with Maxwell Cade[8] that involved a post-hypnotic suggestion, didn't you?

HYNEK: Yes, I talked with Max about finding out whether hypnosis could produce a "genuine" UFO sighting. He agreed to try the experiment. I was to visit Max at his home one Sunday afternoon. He had invited several friends, and he had chosen a man named Derek (I don't remember his last name), mainly because he was a good hypnotic subject and also because Derek didn't believe in UFOs. He was actually somewhat hostile to the idea. Max told me that he had seen

---

[8] British medical scientist who uses hypnosis in his work.

Derek the previous Wednesday and had had a hypnotic session with him; he had given him a hypnotic suggestion that I could shake hands only with my left hand—my right hand was supposed to be injured. And when the afternoon came, sure enough, as I went around the room shaking hands with everyone with my right hand, when I got to Derek, he immediately put out his hand towards my left hand ... picked my left hand.

On that Sunday, we had quite a good session with Derek. I watched Max put him into several stages of deep hypnosis and while he was under, Max suggested to Derek that after we had tea (we were, after all, in Britain!), he would wander out into the garden and see in the sky, coming from the northeast, a very unusual object. He deliberately did not say a UFO because Derek had a block against UFOs. It would be something that he had never seen before, a very strange object.

Derek was duly awakened, and eventually we went down and had tea. While we were having our tea, someone suddenly said, "Where's Derek?" We looked around quickly; Derek had slipped into the garden. And those of us who were in on the experiment went out, as casually as we could, deliberately making a point of not looking up at the sky. We looked at flowers, asked questions about this shrub, and so forth. And all of a sudden Derek pointed to the sky and said, "Hey, look, look. There comes something. Can't you see it, can't you see it?"

We all looked up and said, "Where? Where?" Some of us pretended that we could see it, but Derek got quite disgusted with us when we couldn't see it. He said, "It's just near that cloud now. There! Can't you see it?"

We said, "Not quite. Tell us what it looks like." He said, "It's like a football followed by some golf balls."

By the way, Max had said, in his post-hypnotic suggestion, that he would see it only for a short while. Suddenly Derek said, "It's gone. It's gone behind that cloud." And we all pretended to be excited, ohing and ahing, and I said, "Would you mind drawing a picture of what you saw?" "Oh, not at all." He went inside and drew something rather vague, certainly no details, one squiggly sort of football followed by a few tiny little circles representing the golf balls that were following it. They were dots, however, against the sky background, he said.

Another incident involving hypnosis took place at Ottawa, Canada, where I was a guest on the Kreskin show, the "Amazing Kreskin" as he is called. He pulled rather a fast one on me, I thought. He had me on first to talk about UFOs. He didn't tell me what he was going to do. I noticed he had a TV camera set up in the snow outside. It was a cold winter night in Ottawa. In his rather long preparation for his show, he had selected highly suggestible persons. He had fourteen of them on the stage by the time I got there. He assured everybody that there was no such thing as hypnosis. It was all just suggestion. This probably is because hypnotism is not allowed on Canadian TV. So he had to make that demurrer. But he maintained all along that it's all suggestion and not hypnosis.

He put these people through numerous hypnotic tricks, such as having them hold their arms straight out in front of them and telling them that they wouldn't be able to lower their arms. Saying, "Try it. Lower your arms!" They couldn't. He also had them close their eyes and said, "Now you just can't open your eyes. Try opening your eyes. You can't. Try hard! Real hard!" And they couldn't open their eyes.

All of a sudden he pulled a big switch. He said, "Now we are all going outside!" They all piled outside. As soon as he got them out in the snow, with TV cameras going, he suddenly gave them a key word, a code word which would put them under hypnosis (suggestion!) again. He used that outside, so they snapped back into their suggestible state. Now he looked up in the sky and said, "Now there's a UFO. Do you see it? You do see it? A UFO!" Every damn one of the fourteen saw it! One of them got quite violent. We had to restrain him. He wanted to dash madly back into the studio saying, "We've got to call the police, we've got to call the police!"

We managed to keep him from doing that. Every one of the fourteen saw something, or said they saw something. Of course none of the rest of us did, although we looked very carefully. Quite obviously there was nothing there.

Then Kreskin took them out of the hypnosis and got each one in front of the TV camera and asked them, "Have you ever seen a UFO?" "No." Yet they just said they had seen one a few moments before.

I suggested that this was a fine piece of evidence and that we ought to have them come inside and draw what they had seen. Though that wasn't part of the program, he went along with it. We got them inside. By this time the program was actually over—it was a half-hour program. We got out sheets of paper, and they all drew things, all quite different, all quite vague. No elaborate constructions. Just sort of vague blobs and globes, circles, and so forth. It certainly demonstrates that a person can be made to think that he sees a UFO if deliberately hypnotized! But he is essentially incapable of describing or sketching what he claims to have seen. It seems apparent that any explanation of the UFO phenomenon as the result of hypnosis—self or otherwise—just won't stand up.

5

The Night an Occupant Was Shot

Apparitions and Contactees

In the first four chapters of this book, we have seen cases involving low-flying objects in which occupants were clearly described (in New Guinea). We also discussed the question of witnesses who appear to remember being "abducted" aboard an object and whose description of the event is only possible under hypnosis. We have hinted, too, at even more complex situations—those where the drama involves a direct physical relationship with the occupants. The following case is transcribed from an interview that took place in January 1968, conducted by a U.S. Treasury agent who wishes to remain anonymous. It contains several elements that force us to consider it at the top of the "strangeness" scale. It has never been explained. By its "dreamlike" quality, it reminds the investigator of the Betty Hill case and of the ghost stories we have all read as children. On the other hand, it rests on enough hard data to be taken seriously.

The incident took place on a cold November night in the northern plains of the United States. Four men were driving back from a hunting trip through the snow-covered landscape. In their car was a loaded gun....

  Q. Now, let's see, Mr. H., this was the last part of November 1961 when this took place, is that correct?

  A. That's correct.

  Q. You and Mr. D. were in the back seat asleep while Mr. R. and the driver of the car, Mr. T., were in the front seat, right?

A. Yes.

Q. Just go on from there and tell the story in your own words.

A. We were driving down the highway, going toward home after this hunt. It was raining, sleeting, and the rain was freezing on the windshield, and it made visibility real bad. Anyway, this man who was driving the vehicle was the only one awake if I recall right. The heating system wasn't too good in the car and, therefore, a lot of ice was freezing on the windshield. *He noticed this object in the sky that was coming down. It looked like it was all ablaze, you know, on fire.* So he kept his eye on it and it seemed to come down, oh, probably half a mile to the right of the car and up ahead near the highway.

Q. This is what he told you after he woke all of you up?

A. Right. At this time he nudges the other fellow sitting in the seat with him [Mr. R.] and gets him awake. He gets to see this object coming down also. *It went out of sight, and they thought it was an airplane that went down and crashed.* Now when they got up alongside, having traveled the distance from where they first saw it, they saw the thing sticking in the ground like the tail of an airplane at an eighty-five degree angle.

Q. It was sticking in the ground, tilted somewhat at an angle, is that right?

A. That's right. So they stopped our vehicle. Through this glow they could see it was a shaft or silo-appearing type craft which was sticking in the ground with this glow around it. We had this little hand spotlight that plugged into the cigarette lighter, and they used it to take a better look at the thing. They saw four people standing around the craft, or whatever you want to call it.

Q. How far away from your car at this point would you say this object and the people were?

A. I would estimate it at probably 150 yards; there is a ditch, a fence, the railroad tracks, and it was about fifty yards on the other side of the tracks. When they shined the light on it—this was told to me later—when they shined the light on this vehicle and the four people, well, it was just like there was an explosion, sort of, and uh, everything went out. And that's what we thought it was, an explosion that blew everything up.

Q. Up to this point, you all felt that this probably had been a plane crash or what?

A. Right! Now I'm not quite sure, but if I remember right, this third member of the gang, who was on the right-hand side of the back seat saw the thing come down, too. I know I didn't. I didn't see any of this. I was still asleep in the back seat. I had a parka on and had it pulled up over my head and was all snuggled up, because it was pretty cold with the heater not working and all. They decided to drive on up ahead. They couldn't get across the field at that point, but they wanted to get over there and help these people as best they could. They thought they were burned when the airplane blew up, naturally, so anyway, we drive ahead. And probably half a mile down the road, we turn to the right again, into the field where this craft was, or had been, I guess you would say.

This is the point where they woke me up, and I saw this field was all white with ice from the rain. They were trying to drive across the field. But I guess this little car we were in just didn't have it, and they didn't have too much success. And so they woke me up and told me there'd been an airplane crash, and since I'd been a medic maybe I could help. So they were trying to find the spot where they'd seen this thing sitting, and they couldn't find it. I said, "Well, let's go back to the road. Can you find the spot there where you saw this from?" And Mr. T. said yes, definitely. There was a county road sign and a dead cat lying in the road. So we proceed to turn around, and as we were coming back down the highway we noticed these forms, so we stopped the car. The driver was now on the opposite side, on the same side as the craft, and he saw these forms and the craft again, and he said, "There it is!" At this time I leaned forward and grabbed the light out of his hand and I said, "What do you see?" And I was shining the light up and down on this object that was sitting there. It was like the silo of a barn, like I say, sitting there and—

Q. Was it silver in color you mean?

A. Silver in color. Just as I got the light on it and was moving it up and down, I said, "There's one of them." And the person, or whatever it was, was I'd say about five-feet-five or less and wearing what looked like white coveralls. As far as headgear, I can't remember. I thought about this quite a few times afterward, and I know he had something on, but I don't know what it was. It didn't cover his whole head though. And at this point we were all getting kind of shook up, figuring someone's probably hurt there. We even started to get out of the car to go over there, and I held the light on this person. *At this time, he made this gesture of, well, move back, or get out of here!* Of course, we all noticed this and we couldn't figure it out. If they were hurt, if there had been an airplane crash, they would want help instead of trying to push us away. Unless it was something we weren't supposed to see. Like an Air Force vehicle that was new in design. But we didn't know, so we started talking this thing over and they mentioned to me that there were four of them; I don't know where the other three were. Well, of course, one of the fellows in our crew is a pretty big boy, six-foot-three, 230 pounds, and is also a wrestler. Well, he wants to get out there and see what the hell is going on. We debated this for a few minutes and finally two of us decide to go take a look. The rifle came in here. We had a .22 Hornet, a Winchester Model 43, scoped with a Weaver K-4. At this point I said, "No, no! You can't shoot a farmer!" I said that might be a damned silo out there and somebody walking around it. Of course then the argument starts back and forth.

"I'm Gonna Get It Off My Mind!"

Q. What was the reason for the rifle coming up at this point and why did you all start talking along this line? Did you all at this point have the feeling that it was something other than a plane crash?

A. No, I don't believe we did. Except maybe for one of the individuals who, well, I don't know, because *they were human beings* and I saw one of them. I saw him walk. I saw him move his arm in that "get out of here" signal. I saw him move away.

Q. In other words, the *people* out in the field appeared to be human in all respects, right?

A. Right. Definitely, yes. Now this fellow that was driving the car, he mentions the fact that he'd seen this highway patrolman sitting in a car off the main road in this little town we'd passed through, a town that has only one brick building in it. That's all I remember about the town. I've never been back to try and find the spot or the building. I've thought about it lots of times, but I've never gone back there. Anyway, we decided to drive back, and from the spot on the road where the sign and the dead cat were, take a speedometer reading so we would know how far we'd driven and could find the spot again. So we went back. We explained to this policeman what we'd seen. He wants to know where we'd been. We had been up all that weekend. The night before, we'd driven all night and hunted all day, and come back that evening. So he wanted to know all about this and whether we had been drinking and so forth.

Q. Now, this is on Sunday evening, right?

A. Yes. We all had to go to work the next evening. Anyway, we told the officer, and he looked in our car and saw that we were hunters and there was no liquor of any kind in our car. I'm sure that he could see we hadn't been drinking. He told us, though, that if this was a hoax or a joke, he was going to run us all in. And I remember that this officer did take our names and he did write them down. Anyway, we persuaded this guy to come back to the spot. It was exactly eleven miles from this town. We got to the spot. We stopped, and he pulled in right behind us. We shined the light over toward this spot where we had seen the craft the last time, and there wasn't one person. It wasn't there. Of course, he could see that and we were quite upset. He was too. So we were standing there talking, and we noticed off to the left something like taillights. To me, it looked like a car moving around out there. The officer said he was familiar with the area and there were no roads, nothing but a field out there. So he said, "Come on, let's go out and see what the hell that is." So we went back into the field again, and I think one of us was in the car with him.

Q. It was Mr. D. who got in the police car, right?

A. Yes. I think so. I can't remember too well. As I say, it was a fairly screwed up night. Anyway, they got up right behind this "vehicle" and we pulled in behind them. All of a sudden, it was just like they had caught him, it shut its lights out! When we pulled up, they were pretty excited, the thing had disappeared! We got out of the car, looked around and there were no tracks. It was gone, that's all there is to it, but we all saw it, we don't know what it was or who it was.

Q. This officer, and Mr. D., who was in the car with him, did they pursue this "car" either out in the field or down a road?

A. This was out in the field. There was another country road, a graded road, but the pursuit was in the field.

Q. The lights on the object appeared to look like the taillights on a car, and they just went off all of a sudden?

A. Right, they just went off, that was all there was to it.

Q. There were no tracks left in the field or any indications that it had even been there?

A. Right, no tracks or anything else. Like I say, it was raining, sleeting, and the ground was still pretty muddy, and I feel that it should have left tracks; it was deep mud, just about up to the hubcaps.

Q. At this point then, the officer decided to go on back to the town?

A. Right.

Q. Did he mention anything as to what he felt or thought at the time as to what this may have been?

A. He thought it was awfully damned strange, and he couldn't figure it out. We didn't know what the hell it was; it was quite an experience. He didn't know what to think of it, and we talked for quite a while, and he finally said, "Well, it's one of those things," and we took off, went down back to the road. That's the last time I ever saw the man.

Q. Okay. After he left, what did you guys do—get back in your car and start back or what?

A. We started back home, and this is where . . . I just made a phone call before I came up here and I talked to one of the fellows involved, and he told me I was crazy as hell for talking to you, that we'd all made a pact that we'd never mention this part of it to anyone, and that I was a damned fool for doing this. But I've been living with this thing for over six years now, and *I'm gonna get it off my mind!* We were proceeding down the highway towards home, and I imagine we went about two miles down the highway, and *this object appeared again.* So we stopped the car and noticed the thing was coming down.

Q. You saw the object coming down this time yourself, correct?

A. I saw it coming down this time. It had the glow and was really quite a sight to see.

Q. Did this "glow" appear to be flame coming out of it, like an exhaust of some type?

A. No. It was more like a glow, if you watch an arc welder—

Q. White glow?

A. Well no, it was more of a red. Red with, I don't know, the outside was a real sharp glow, like a white or blue, real hot looking; it sat down, just like a chopper would sit down, nice and gentle, no problem at all; this was under 150 yards away, I know it, and they got out again. There were two individuals this time; they had the same type clothing.

Q. Did you see them actually get out of the thing? How did you first see them?

A. *We did not see them get out.* How they got out of the machine I don't know, it was just glowing on the bottom, still had the glow to it. Again we whipped out the light, we talked things over—what we were going to do. Two of us got out of the car at this point. Now there were the same railroad tracks, and I think the craft was just a little closer at this time, and there was a knoll to the right and one to the left and in the center this saddle or hollowed out spot,

I'd say it was probably sixty to seventy feet lower than the rest of the terrain around there, maybe not that much; anyway, one of us from the car went to the left and one went to the right. The one to the right was carrying this .22 Hornet rifle with the scope. Now we got out there, and I don't know just what we were thinking of, but the two men left in the car were holding the light on the "two forms from the vehicle." They were pretty close together, I'd say about ten yards apart, one was a little forward of the other one.

Q. They were just standing there watching you?

A. They were just standing watching; so by this time, the fellow on the right knoll dropped down on his knees, then down to a prone position with the rifle. This person looked in the scope and examined the person from the knees on up. Like I said, I don't know what kind of headgear, this is the thing that really sticks in my mind.

"Have You Ever Shot at Night with a Scope?"

Q. Now the light enabling the fellows on the knoll to see the two forms was coming from the hand spot held by the two men back in the car?

A. Yes, and if you've ever shot at night with a scope and with lights, no matter how dark it is off to the side, you look at the object that you're going to fire at and you can see it real well. Anyway, *at this point there was a shot fired, it hit in the right shoulder of one of the "forms,"* high in the right shoulder. When the individual was struck, he spun around, down to his knees and then he got up with the other guy's assistance, and he looked over and said, or hollered, "Now what the hell did you do that for?" Everyone at this point was pretty well shook up, and there was a scramble back to the car. And well that's the story.

Q. Do you recall at this point what happened concerning these guys getting out of there? Did you see them get back in the object and the object lift up in the air again and take off or what?

A. No. No. No recollection.

Q. No recollection on anyone's part?

A. No. I guess we were just concerned with getting out of there and never looked back.

Q. Okay. Now as far as the shot that was fired, you mentioned that it hit high and in the right shoulder. Was there an impact that could be heard, I mean beside the fact that the guy spun around? Could you hear the impact?

A. Yes, it was an impact, thud sound, just like shooting any animal. You can always tell when you hit.

Q. Did you see any blood?

A. No. The person who fired the shot never did recall seeing any blood. Everyone was too shook up at this point to notice anything from then on. We just wanted to get the hell out of there. But the shot definitely hit something that was solid, it wasn't just a bunch of cloth floating in the air.

Q. Now, after the shooting incident and the two men had returned to the car from their positions on the knolls, what was said at the car?

A. I don't recall what was said.

Q. Didn't you mention something to me to the effect that one fellow said let's get the hell out of here 'cause he just shot a man, or something like that?

A. And the two men in the car said the rifle had never been taken from the car. Yes, I remember, and someone was sure screwed up.

Q. Apparently the two guys in the car, or I should say, were the two fellows still in the car even aware that the other two had been gone from the car?

A. Well, this is really hard. This is one of the arguments we've had all this time, two of us definitely know there was a shot fired, and two say, "You're crazier than hell." This is what they'll maintain if you talk to these people; I don't know why. Just like the person who fired the shot; he's a fairly good shot, and would never shoot at anything he wasn't sure of or unless he knew what it was, and why the trigger went off I don't know, or why the rifle was even taken out of the car I don't know.

Q. After these two individuals returned to the car, and you had your discussion about the rifle never being out of the car, then what happened? Did you start on your way back or what?

A. Yes, we came on back. We got home and it was pretty late.

The Loss of Time

Q. You did mention to me one time before that all of you had more or less agreed after you returned that *there was a certain period of time that you had lost somewhere enroute, and you couldn't figure out what it was or where?*

A. That is right.

Q. What made you aware that there was a period of time missing in there someplace?

A. From the time that we started this trip, we figured it was earlier, and by the time we got home it was just daylight; the wives were all sitting and waiting for us. We all knew it took too long for us to come back; we hadn't spent that much time chasing this thing around. All of us knew and had the feeling that there was something missing there, *and to this day we don't know what it was.*

Q. Concerning the shot that was fired, you say the one individual helped the other one to his feet after he had been hit. Did you see what they did then, did they retreat to this vehicle?

A. No, I think we ran too fast to notice that.

Q. Was the object still glowing at the time, did you notice?

A. To tell you the truth, I never even looked back at that thing again after watching these people; I never even looked back to see if this thing was still there. I mean this is the thing I don't know.

Q. All right, now you all have returned home, and of course the pact was made not to discuss this with anyone; what happened to you the next day concerning some people who came to see you where you work?

The Intimidation

A. I was called by my supervisor to go downstairs to see these individuals, and at the time they introduced themselves (name only) and asked me if the event that I had seen the night before was true; I said yes, and *they wanted to know the type of clothing I was wearing;* they asked me a few questions about this object we had seen and most of their questions were just like they knew what we had seen and wanted to find out how much we had seen; anyway they asked about the clothing I was wearing.

Q. When they asked you this, did they come right out and say, "We heard," or, "We received a report," or what?

A. They *knew my name, they knew where I worked,* and I just took it for granted that they had this report from this law officer down there, and they were just checking this thing out to see if we'd seen it. Of course, I was still pretty shaky over the events that had happened the night before, especially the last part of it, but the law officer didn't know anything about that part of it, he'd already gone back to town by that time, so I figure, well, this is the U.S. Air Force, and as we had given this officer our names, I figured they knew just who the people were that shot the Air Force people. So they asked me the type of gear I was wearing; I told them hunting clothes. They asked me if I'd gotten out of the car in the field where the first sighting was made, and they're strictly talking about the first sighting. I told them yes, and the conversation went on and they asked about footgear. I told them I had on hunting boots.

Q. Wasn't this after you had accompanied them back out to your house?

A. Yes. They took me to my home to look at my hunting gear and my boots. They never did say why they wanted to see these things. The only thing they asked was whether I'd gotten out of the car into the muddy field, and I said yes, and that's when I showed them the boots I'd had on. They asked if I had any other type of boots. I said yes, I had a pair of wader type boots. Of course, I told them I hadn't been wearing them, but they wanted to see them anyway. So I produced them. They thought on it a little more and said, "Okay, that's enough for now," and they left—real odd [shakes his head].

Q. When you say they left, you mean you were left at your house?

A. Yes, they just got in their car and left. I had ridden out to my home in their car, so I had to call a cab to get back to work. My wife was at work. And that's the last and only time anyone had ever questioned me about this sighting until you came around.

Q. When they first approached you at work, did they ever show any credentials or represent themselves as being with the government or what?

A. No.

Q. They said the information had been reported to them, is that correct?

A. Yes, they said they had received this information in a report, and now if you can just stop and think of my reactions. This was the next day after, just about noon. They were well-groomed men, *they were official looking*, just like they had a job to do; I mean they were guys that acted real businesslike.

Q. Were they in civilian clothes?

A. All in civilian clothes, I don't remember what they had on. I remember they all had ties on, and as I say well groomed, and the same type of people you and I are, no different. In fact, I figured they were intelligence officers of some type, and I'm pretty well shook up at this point to have these people come right out like this, after what had happened the night before; therefore I just said yes and no to everything they asked and went along with them. After all, I figured they had me, and probably knew what they were looking for.

Q. Do you recall what type of car they were driving?

A. Yes it was ... no, I don't either; I think it was a white car. I remember that it was a model of the year, I think it was a '61 Plymouth; I'm not sure of that. There was nothing fancy about the car. It was just strictly a mainline type of car with no padded dash or anything like that. When I got in the car, I got in front with the one individual and the other two got in the back.

Q. Did you notice any license plates or anything like that?

A. No, I was pretty shook up, I just got into the car with them; they had it right out in front. They seemed to know everything about me, where I worked, my name, and everything else.

Q. After the incident at the house, am I correct in recalling that you said there was some mention made by them in the form of a warning not to talk about this to anyone?

A. Yes, in fact the one guy, the one who did all the questioning, said we want to thank you, and he called me by my name, for your cooperation, and he said we want you to keep this quiet, *"you'd better not say anything about this to anyone from now on,"* and I assured him I wouldn't. So they got in their car. I then got pretty upset as they'd left me at home with no way back to work.

Q. Have you ever been approached since by these men or anyone like them?

A. Never, never.

Q. While they were questioning you, did they ever make any mention to you about the shooting incident out there?

A. Never once. That was never mentioned at all.

Q. Then it appeared that they were only aware or interested in the first part of the sighting?

A. As I told you, they never asked anything about the shooting and all their questions were concerned entirely with the first part of the sighting. I think they probably knew more than they said, but I don't know. This has been on my mind for quite a while and, uh, it isn't so bad now, but the first year after that, every time I'd get a knock on the door I figured it would be those boys wanting to know more about what happened!

Q. Is there anything that has happened to you since that time that you feel might be related to this whole incident in some way?

A. Well, this I'd better not say, you know. No, I'm not going to comment on that.

What Apparitions Do

The psychological impact of such a case is enormous, as can be discovered from the discussion between the questioner and this witness, for whom the incident still has a dreamlike quality many years after the sighting. The loss of time is reminiscent of the Betty Hill case examined in the previous chapter. The argument among the men as to whether or not the rifle had been taken out of the car is amazing. These witnesses had in fact lost their sense of reality. They had gone beyond the edge, entered an alternate world. The visitation by official-looking men the next day made matters even worse. As we say in Chapter 2, this is not an isolated case.

What did these men see? What is the nature of an "apparition"? These questions formed the basis for the next discussion between the authors, with psychologist Dr. Arthur Hastings again sitting in. Discussing this case, one of us raised the question of its physical nature.

HASTINGS: Let me comment on apparitions briefly. You're right that most apparitions are very real. In fact, most people don't know they're [seeing] apparitions until they've disappeared or walked through a wall! Sometimes, however, they don't have any feet. Did you know that?

HYNEK: No, I didn't know that.

HASTINGS: Sometimes a person will be asked, how did it react? What was it wearing on its feet? And the person will think for a moment and suddenly say, "It didn't have any feet!"

HYNEK: But on the other hand, a British scientist named Maxwell Cade[1] told me that in post-hypnotic suggestions, you can say that at this party Miss So-and-So will be there and this person sees her there. And when he is asked later what she was wearing, he will make up what kind of dress she had on and, on a man, what color tie he had—they improvise. So I don't see why they wouldn't improvise here too, unless it struck them very hard that it didn't have any feet.

HASTINGS: I'm assuming they're seeing something real, something that's genuinely there. In hypnosis, you can't say they're making up that because for all you know they saw a person standing there wearing these kinds of clothes. When I ask people to hallucinate little black dogs, so far as I can tell they see a little black dog with a head and a tail.

HYNEK: Yes, but did you ask them, did it have a collar? Will they improvise a collar?

---

[1] Maxwell Cade is the scientist with whom Dr. Hynek conducted the hypnosis experiment related in Chapter 4.

HASTINGS: How do you know they're improvising if they say it has a collar? Now Tyrrell, a leading British psychical researcher, thinks apparitions are not things that are physically there in any sort of sub-matter form but are rather created by stimulation of the sensory system inside the person.

HYNEK: Well, if they are apparitions, they're apparitions that can produce tangible effects.

HASTINGS: The apparitions in the classical occult literature do not produce physical effects. They don't drop handkerchiefs or knock over things with their elbows.

VALLÉE: They don't usually cause macroscopic effects, although they may wither flowers or leave a strange odor or . . .

HASTINGS: Change the temperature . . .

VALLÉE: Cause a draft of air . . .

HYNEK: Well, I've got to tell you now the story I was going to tell you at dinner the other night. I have a friend who has a strong interest in the Hopi Indians and the Kachina dolls. The Kachina dolls are supposed to be representatives of the sky people, the gods of outer space. When I was in Batineau, North Dakota, in this very isolated place, I met a historian of Indian cultures who started to tell me some stories of the Indians. All tribes have these stories of sky people, not only in the Hopi culture but also the Sioux and others. The Sioux, for instance, state that when the sky people left, *they changed themselves into arrows and went off into the sky.* Well, so did Neil Armstrong change himself into an arrow when he got in his rocket and went off!

We got talking about medicine men; he was telling me that he'd been accepted into the Indian culture. He has done the sun dances with them, and he told me of two instances of clairvoyance. In one case, an Indian named Red Horn was one of three young bucks who were drowned. They found the bodies of two of them, but Red Horn was not found and so they did the windshaking dance (they have another name, but this guy calls it that). He said there is a long, involved dance and ceremony; when the Great Spirit comes, there is a very real wind that just zooms through the village and slams doors, and the medicine man says, "The Great Spirit is here!" One sideline: for some reason, some of the younger members of the tribe that were going to perform this dance wanted this Catholic priest to witness the thing. And when the Great Spirit came, the Catholic priest cried out in pain. Later, he said he felt that every bone in his body was broken, and they just laughed and said, "Well, that's what the Great Spirit does to those who don't believe!" There was no permanent injury, but it knocked the hell out of him. And then after a number of minutes, the Great Spirit left, there was a wind in the other direction, and the medicine man said, "The Great Spirit is gone!"

The instructions that the medicine man gave to the family of Red Horn were to watch for three large birds. They were to follow these birds. At a certain point, two birds would turn off and one would go straight; they were to follow the one that went straight and they would find the body of Red Horn. So they alerted everybody to watch for the birds, and they came; two did turn off and one went straight, and suddenly a shot rang out and then one fell down. The Indians didn't shoot it. Somebody else, nobody knows who did. But somebody shot it and it came down and right under where it fell in the water they found Red Horn's body.

Another case he told me was where a young Indian had had a very valuable saddle stolen. So they again had a ceremony and finally the medicine man said that on a certain day (and he specified the day) this buck was to go out along the river until he saw a coyote sitting on a bluff. He was to dig under that bluff right where the coyote had been sitting, and he would find his saddle, and he did. Now, of course, those things are pretty hard to believe in a sense; on the other hand, how do we know?

HASTINGS: Yes, well, they're certainly consistent. It is interesting that our clairvoyants tend to describe the locale "there's a bluff with a certain kind of rock under which"—but to follow three birds, that's something else.

HYNEK: Well, this is very much like Carlos Castaneda's *Journey to Ixtlan*, where he "has the power" or he "doesn't have the power," and he is always looking for signs, for a swallow to fly a certain way, and that sort of thing.

VALLÉE: I have heard of a local healer in Normandy who is also a clairvoyant. Someone lost a sheep and went to see him. He described the locale by remote viewing; he said he saw a river, a bend in a river, and some trees and a depression, and he "saw" the sheep there. That wasn't in the area that the owner of the sheep knew, so he asked around if anyone knew any area like that, and the people said there was such a place ten miles in such a direction; he went there and found the sheep. But it was a case of remote viewing again. There was no "sign."

One thing that bothers me whenever this subject of links with the occult comes up is a curious tendency on the part of even the most knowledgeable of those people to reject UFOs just as scientists do. For instance, you approach the subject of UFOs with people who are into the cabala or astrology, alchemy, spiritualism, or other esoteric topics. These are very well-established disciplines with certain hierarchies of subjects. But whenever you mention UFOs, they just don't fit into any of their existing categories; there is no occult school that has a tradition of UFOs.

They are just as baffled about it as the rationalists, and they tend to reject it with the same kind of skepticism; in other words, you find the same kind of skepticism in the high priests of the occult as you find in the high priests of science.

HASTINGS: That's quite true, absolutely true. Now why is that? The behavior of the UFOs, the technology, might have some similarities. But the behavior, the intelligence, and the occupants absolutely have no place in any of the occult traditions.

HYNEK: But what about the humanoids? If you're going to say they resemble anything, it seems to me if I were pressed to that, I would say they resemble large versions of the Little People, or sometimes the things that are called Elementals in the occult literature.

VALLÉE: Yes, but there really is no occult tradition about contact with the Little People. There are Rosicrucian theories of contact with Elementals, but Elementals and the Little People (the Elves and Leprechauns of Celtic lore) are two different things.

HASTINGS: That's true, sometimes the Little People are seen, but that's about it.

VALLÉE: A good example is that celebrated magician Aleister Crowley. One day in Switzerland he saw a little creature and it was clear to him it was not an Elemental, and he had no idea what it was.

HYNEK: How are we defining Elemental now? In most of the traditional occult literature, Elementals are said to be creatures in the "etheric realm," and/or astral domain. An Elemental frequently can be created by a person and infused with his own emotion and becomes an actual creation of this person; whereas the Little People supposedly represent another scheme of evolution, another life wave that is not connected with the human race. Some occult literature terms these— salamanders, undines, sylphs—Elementals, too. Confusing! Well, there we are! What a strange turn the conversation has taken!

The Occult Connection

HASTINGS: I think that's an important point Jacques brings out, that the occult seems really quite different from the UFO phenomenon.

HYNEK: Yes, that's a good point to bring up. The booksellers insist on putting UFO books on the occult shelves. And you say, "Well, doggone it, there's no obvious connection."

VALLÉE: Well, maybe there is, but it's certainly not clear to what extent, if there is a connection.... To me, that's an indication of how important and central the subject is, that it doesn't fit into any of the existing categories. When *Anatomy of a*

Phenomenon came out, for instance, I would see it in the bookstores under science; but sometimes it was under science fiction, sometimes it was under occult, and sometimes it was under religion.

HASTINGS: Okay, look at what we have now. I just see a couple of connections. We have one connection of one type of humanoids with the fairies, with the Little People.

We have a connection of a different type of humanoid with the gods from outer space, with legends of gods, space people. We have another connection which seems to be growing with some kind of robot creature that we don't have in our history except for recent science fiction. But both are quite different traditions. Space people are not Little People. Those are separate traditions.

HYNEK: The large ones are generally more elegant, more stately, and reportedly more helpful.

VALLÉE: In other words, I am not sure whether it is cultural anthropology or psychology, but we have two categories of occupants: the outside alien and the earthbound alien. And the Little People are definitely earthbound. They live in caves, they have oversized heads, they seem to be accustomed to a dark environment, which is consistent with the idea that in all traditions they come out at night, they are mostly night people . . . large eyes, very clever, psychic as hell, and living in caves. Difficult to detect. The contact is on their terms. Then, on the other hand, you have the outside alien who is described as a god from space, a creature from space. The problem is that in both cases, there is contact with man, and in many UFO cases, the occupants themselves are, in fact, men. In the case where the occupant was shot, they were described as ordinary men in white clothing. And in the Eagle River Case,[2] they were ordinary men.

HYNEK: They were described as "short Italians"!

VALLÉE: Short Italians, yes, but wearing a black uniform with a stripe down the side, white stripe, like a ski suit!

HASTINGS: Also, of course, the people we are seeing now are behaving like space gods. In the religious traditions, those have been pretty clear. Creatures with lots of light, and dressed up for the occasion!

HYNEK: Yes, they're not creatures of the dark.

HASTINGS: Nor are they Italians.

VALLÉE: My daughter Catherine, when she was about three, had fantasies about two categories of creatures. At least I *thought* those were fantasies! The creatures

---

[2] See "The Landing at Eagle River" on page 127.

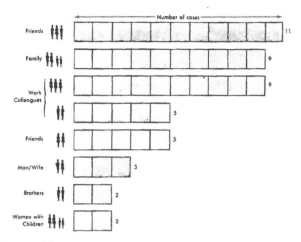

Figure 16 Sociology of the UFO phenomenon. Composition of groups of witnesses in one hundred UFO landings observed in Spain, based on a study by Vallée and Ballester-Olmos. This distribution indicates that witnesses of close encounters with UFOs are typically with friends, with their families, or with work colleagues at the time of the sighting—hardly the expected social situation for a hoax.

were called "Meu," and they had different colors and different attributes. If I remember correctly, there was *Le Meu Noir* and *Le Meu Rouge*. The Black Meu and the Red Meu. The Black Meu had definite characteristics. He could jump, he could talk, he could do almost anything. He was a monster-type creature, playing pranks and things like that. But *Le Meu Rouge* was a completely different kind of entity: you couldn't see him, you couldn't touch him. He had no feet. He had only negative attributes, he was defined only in negative terms. There was nothing positive you could say about him.

It seems to me that in many of those sightings, in the descriptions of occupants and so forth, you can find the same kind of quality you can find in children's stories describing something like this. Perhaps we should just bring out the fact that very young children fantasize about certain creatures that are broken into distinct categories. You see, she is six now and it's all over. She doesn't see them anymore.

HYNEK: Since she is right here at home, why don't you ask her about them?

VALLÉE: [Calling his daughter] Cathy ... I was telling Dr. Hynek about the Meu. What color was it?

CATHY: Oh, red. One was black, too.

VALLÉE: How big was the black one?

CATHY: Oh, it was sort of small.

VALLÉE: How small?

HASTINGS: As tall as you are now, or shorter?

CATHY: The red one was about like that.

VALLÉE: Does he have feet?

CATHY: No, he just sneaks around.

VALLÉE: Sneaks around? Do you still see him? When did you last see him?

CATHY: Oh, a long time ago.

VALLÉE: Can you draw him?

CATHY: Sure, I think.

VALLÉE: Let's draw him. Got a red pencil? She can draw the red one.

CATHY: There, like that.

VALLÉE: Does he have eyes?

CATHY: Just sort of ...

VALLÉE: He doesn't *have* to have eyes. What else does he have?

CATHY: Well, inside he's red.

VALLÉE: He's all red inside. When did you see him? In the day or at night?

CATHY: At night.

VALLÉE: Does he come out in the day?

CATHY: No.

VALLÉE: Where is he during the day?

CATHY: Oh, playing with ghosts.

VALLÉE: Playing with ghosts?

CATHY: Yes, *of course!*

VALLÉE: And how does he go around?

CATHY: Oh, he sneaks around very slow.

VALLÉE: Do you see him inside the house sometimes?

CATHY: No, I only see him outside.

VALLÉE: Outside? How can you see him outside at night?

CATHY: Well, first my Mom forgot to close my curtain.

VALLÉE: And then he was outside?

CATHY: At night he was outside.

VALLÉE: What did he do?

CATHY: Oh, nothing. He went "moo" sort of like a cow.

VALLÉE: And where does he live?

CATHY: Probably with the ghost.

VALLÉE: In a house?

CATHY: Probably in a haunted house.

VALLÉE: And what about the black one, can you draw the black one?

CATHY: The black one was sort of small like this . . .

VALLÉE: And what does he do?

CATHY: Well, I just thought of him at night walking in a movie.

VALLÉE: What does he do during the day?

CATHY: Probably lives with the red one.

VALLÉE: Does he talk with you?

CATHY: No.

VALLÉE: If I go out, can I see him?

CATHY: No.

VALLÉE: Why not?

CATHY: Oh, I don't know, in my head.

HYNEK: How many do you see at one time?

CATHY: I saw two of them, one in my head and one outside.

HASTINGS: Do you ever see the black one outside, outside of your head?

CATHY: No.

VALLÉE: Do they have teeth?

CATHY: No, I don't think so.

VALLÉE: What else do they do? Or how fast do they go when they go?

CATHY: They go very slow.

VALLÉE: Why do they come near the house?

CATHY: I don't know. They probably want to stay with the ghost.

VALLÉE: How do they go around?

CATHY: They sneak on their knees.

VALLÉE: I don't see their knees.

CATHY: So?

VALLÉE: What do they eat?

CATHY: I think they eat nothing.

VALLÉE: And how do they talk?

CATHY: They don't talk.

VALLÉE: Can they grow?

CATHY: No. They stay small.

VALLÉE: The red one never says anything to the black one?

CATHY: No.

What Children See

VALLÉE: That's the kind of thing that we educate out of children, the ability to see those things. Now she says one is in her head. At the time, it wasn't so clear; it was very real to her.

HASTINGS: One of my friends has a child who is having the same kind of experience, a red light with a man under it. Didn't Cathy say once that her creature had eyes all over it? That's a very distinct thing because in the Bible, in the Revelation of St. John the Divine, there is a description of one of the cherubim with eyes all over it, and in Doré's illustration of that, there is a picture of this creature and it has eyes all over it! And when you look at it, you realize what he meant was that all parts of it were conscious.

HYNEK: I've been asking various Jesuits and other types what the status of angels is, and they don't know, they aren't interested. And yet, "an angel of the Lord came unto Mary," etc. What status do angels have? I tell them, for heaven's sake, it's part of your theology!

HASTINGS: They're literary figures, of course!

HYNEK: Well, according to the Rosicrucians, the angels are the next life wave above us. They started previous to us, they have etheric and astral bodies but simply don't have physical bodies and they'll never have one.

VALLÉE: Except that they cohabited with the daughters of men, of course, and found them exotic!

HYNEK: Genesis 6 says that the sons of God married the daughters of men. The sons of God are not exactly the same thing as angels.

HASTINGS: The sons of heaven.

HYNEK: The sons of heaven, is that what they say?

HASTINGS: It's pretty clear that the way the occupants are behaving isn't consistent with any previous tradition or descriptions ... it might be a gamesmanship thing. They might be doing it deliberately.

VALLÉE: Well, we have discussed a lot of material on apparitions here. We have even interviewed a witness here! This is important because this is the thing we educate out of children and there is no evidence that it doesn't come back. For example, if a witness is exposed suddenly to a very traumatic experience, to something that breaks the continuity of his reality as was the case with Mr. H. in the incident we quoted, it's not clear that this kind of frame of mind doesn't take over. Mr. H. might have regressed to it, he might have lost his adult sense of reality.

HASTINGS: Take, for example, your question, "Where are its eyes?" Maybe it didn't have any. How does it get around? What does it do? You see, if we program people to think in terms of those categories, then it's very easy to simply project a hallucination to include those, of necessity. And if we educate it out ... Still, though, you would think that most people, if they saw something that had different characteristics, and they projected those characteristics, you would be able to detect certain kinds of psychological reaction to that necessity. I don't think that would be an automatic unconscious process. There would be initial disorientation, some evidence of an altered psychological process. It's quite possible that the entities themselves could project into our mind before there is any awareness on our part that they should be seen in this way. I don't think that's beyond our imaginations, beyond our technique. Carlos Castenada talks about "Allies," which he says to most people look like people, although the *colors* are different. When you're in the sorcerer's world, they look like triangles drifting downwards or something like this. Might children see them until they are too old?

The Landing at Eagle River

VALLÉE: What about situations like the Eagle River case that was mentioned earlier, where the occupants were again ordinary men? You remember, the man in Wisconsin with the pancakes?

HYNEK: Oh, good old Joe Simonton!

VALLÉE: Joe Simonton. Now he wasn't in the same psychological category as Adamski.[3]

---

[3] For a discussion of the Adamski case, see Chapter 6.

HYNEK: No, I'm beginning to think more of Joe Simonton than I did at that time.

VALLÉE: I think he is telling the truth, frankly.

HYNEK: So do I! He was. The things he said fit now, whereas they didn't fit, at least to me they didn't fit at that time. I thought it was just sheer nonsense. You know the story of Joe Simonton?

Well, suddenly the Air Force hears that a UFO had allegedly landed in Eagle River, Wisconsin, and some pancakes had been given to this guy by the occupants. I went up there and talked with him and took pictures and so forth. First of all, I was not at all impressed with him personally or with his surroundings. He lived by himself, he'd been divorced, he lived in a sort of a shack on the outskirts of town, and there was nothing to give you a feeling of confidence. This man could have been a wino; the yard was sort of unkempt, bottles and so forth, not wine bottles necessarily, but just untidy.

His story was that he was having breakfast one morning, and he heard a whining noise outside; he looked out the window, and there was this silvery ship descending. It was hovering; it didn't land, just hovered in the backyard and, of course, he went out to see what was happening. The door opened and a creature beckoned to him—and then as he got close, one of them handed down the most beautiful thermos jug he'd ever seen. He said that the creature didn't talk, but indicated by sign language: water, you know. So he got the idea. He went inside and filled it with water. He brought it out, and he indicated now you can drink! But they must have misunderstood him, because they thought he wanted something to eat, so they handed him these pancakes.

I kept a sample of the pancake and took it back to Dayton. My interpretation at that time was that he'd been having pancakes himself for breakfast and had suddenly had a waking dream or what is known in psychology, I believe, as an isolation hypnosis or isolation delusion, and if he'd had his family with him or other people around, it would have been quite different. A delusion could then have been ruled out. That's why I don't like single witness cases—as they used to say in Roman law, "One witness is no witness."

Then he said it just took off and in two or three seconds it was gone, and I said, "No sonic boom?" No. The trees waved a little bit, but no. Well, I just put it down as sheer delusion at the time, but hell, certain little things hit—no nuts and bolts and no rivets—everything very smooth—rapid disappearance, no sonic boom, and the trees being disturbed. I don't know. He certainly wasn't reading any UFO literature.

HASTINGS: Well, what about the pancakes?

HYNEK: Those were examined and were found to be ordinary grain pancakes.

HASTINGS: I meant, the Air Force wasn't able to say, "Well, this is processed pancake batter from Aunt Jemima's!"

HYNEK: They couldn't tell in great detail, although I think they said it was wheatgerm pancake. [Laughter.] Well, you wouldn't get anywhere using a story like that, and I wouldn't use it. First of all, on the general grounds of a single witness—isolated.

VALLÉE: We disagree about that, as you know.

HYNEK: Yes, we do, this is an honest disagreement. I recognize this point, but I think you misunderstand me. I may believe a single witness. But I think there is little positive value in presenting the case outside, because they'll say, it's just a single witness—he might have been lying.

HASTINGS: Isn't there a fairy bread precedent for that, if the fairy is giving food? Of course, then it's not just an ordinary pancake.

VALLÉE: The exchange of food in fairy lore is a very common symbolic gesture, and with Elves, it's a consistent way of making contact—they often give you pancakes.

HASTINGS: In Tolkien's *Ring* books, the fairies give flat pancakelike objects.

VALLÉE: Well, in Irish poetry, they "live on crispy pancakes and yellow tide foam"!

HASTINGS: Maybe they did come from Aunt Jemima! I don't know, they have to get it from somewhere, maybe steal it from the local warehouse?

HYNEK: The only thing that I'm uneasy about, and I expect the two of you are also, is that we recognize that the subject is much more complex than we can present. Jacques has called this the *Magonia Syndrome*. The whole craziness of the thing, the whole absurdity—it's another world, another realm, that seems to have some interlocking with ours, and what we're describing here is just that interlocking.

VALLÉE: But that's what we have to work with.

HYNEK: Of course, that's the only place you can start; the only way you can start to work at it is to examine the interface. You can't start with the other, because you would be lost. Now, Jacques attempts some of that in the direction of the psychic phenomena.[4] It seems that these creatures, like the Pascagoula ones, certainly don't resemble the products of higher evolution as we conceive it. Who would think of a clawed creature coming down and being a representative of a very advanced technology? It just doesn't fit!

VALLÉE: The Blob from outer space!

HASTINGS: ... it's never some guy who walks out, other than in *The Day the Earth Stood Still,* where it was Michael Rennie. It's usually the wildest thing you can think of; it scares the hell out of you.

---

[4] See "The Psychic Component" by J. Vallée. *Psychic Magazine,* Feb. 1974.

HYNEK: It may be part of the whole plan! Perhaps there is a UFO fifth column...

VALLÉE: To purposely mislead us that way...

HASTINGS: In one of my classes, a science-fiction class, I did a fantasy exercise. I had everybody close their eyes and imagine they were going to get in touch with an extraterrestrial creature—so I had them imagine a television screen which was connected to this other entity; turn it on, tune it in, and it gradually became clear. Then I asked them to see what the person or the entity looked like, and they had a chance to get a message from it and to give it a message and then the signal was lost. It took just a few minutes to do that, and then I inquired briefly some of the characteristics. Two or three of them saw it like Mr. Spock (on *Star Trek*), a couple of people saw jellyfish-like creatures, some saw colors, two people out of the fifteen or eighteen saw Richard Nixon, and the messages that they communicated to the creatures—at least the ones that were nonterrestrial—were things like peace, or we welcome you and we want to be friends, things like that; none of this 2 + 2 = 4, but it was an emotional message.

HYNEK: Your reference to "emotional" jogged my memory to way back. I once asked Struve,[5] riding from Chicago to Yerkes Observatory, when I was a graduate student, "I wonder what the human race would be like if we had developed emotionally rather than intellectually? If our whole emphasis had been on emotional development and had developed into feelings rather than trying to probe the chemical nature of the universe." *"Oh,"* he snapped, *"it would be much better if you didn't think of such things!"*

HASTINGS: Really, immediate judgmental reaction.

HYNEK: But you see, if you had a civilization that had grown up under totally different angles....

How Fast Does Thought Travel?

HASTINGS: Does NASA have a briefing or contingency plan set up for their astronauts should they have a face-to-face encounter with alien "occupants"?

VALLÉE: I don't know. Certainly, in the cases when they described seeing UFOs, they didn't seem to have any preset thing to do—no special code that they set up—no special program; they seemed to react very much in surprise.

HYNEK: Neil Armstrong has told me NASA had no such plan, but some institutions have done studies, either Rand or Brookings wrote a report, quite some time ago. Now, when talking about "UFOs and Society" we should bring in the fact that the

---

[5] Dr. Otto Struve, one of the nation's outstanding astronomers, then director of Yerkes Observatory.

idea of UFOs is sort of expanding our consciousness, but so do the recent findings in astronomy. When I was a student, as a matter of fact, it was perfectly permissible for us to think that we were unique in the universe. Because, in those days, they thought the solar system was formed by two stars sideswiping each other—pulling material out. Well, we discovered that, first of all, physically, that wouldn't happen; that condensing wouldn't happen. And secondly, the chances of it are so remote, one chance in ten to the fifteenth or something, that it's virtually impossible.

Today, because nuclear physics has changed our entire view on stellar evolution, it now becomes as likely for a star of the class to which the sun belongs to have planets as it would be for an oak tree to have acorns, you see! It's part of the natural evolution of most single stars; in which case, in our own galaxy, there may be, well, since our galaxy has 150 or more billion stars, it would be anyway, ten or twenty billion that could very well have planets.

Then, let's go down very roughly and simply say, if only one out of ten stars in the galaxy has planets and if out of those, only one out of ten ever develops intelligent life and, out of those, only one out of ten developed into the nuclear age and space age, then still, in our own galaxy, there must be some *ten million civilizations like ours!* Well how many discussions like this are going on at this moment, somewhere in the galaxy? But there is this paradox, or enigma at any rate, that while on one hand the astronomer says it is entirely likely that we have life elsewhere, it does not necessarily follow that that particular life is what causes UFOs. We must be careful to distinguish between "extraterrestrial" life and "alien" life. UFO actions do bespeak some sort of intelligence, but is that intelligence from way out in space or much closer in?

VALLÉE: People are always talking about detecting these civilizations by radio. But it would be unlikely that civilizations in other parts of the galaxy would just happen to be using the same communication concepts.

HYNEK: Good point.

VALLÉE: Because there are few stars as close as ten light years from us, and many more as you go farther away, the most likely distance for a civilization communicating with us would appear to be ten to twenty light years, because beyond that distance their chances of detecting the Earth drop rapidly. That means that the signal would take some fifteen years to come to us. Now, if we look at the evolution of our radio technology in fifteen years, we find that with the equipment of fifteen years ago, we would not be able to detect some of the signals we're using today, especially in communication among computers. Okay, so over the transmission time of the signal, the information transmission technology changes in such a way that the receiving instrumentation is obsolete!

HYNEK: Also this gap, saying hello and waiting years, as somebody said, it's the "long-since-dead talking to the not-yet born" situation. There's got to be some better way of contacting other civilizations!

HASTINGS: It is likely that there is a far better way of doing it.

HYNEK: Of course, the relativists and the skeptics come in right away. Carl Sagan has pointed out that "the average distance between the stars in our galaxy is a few light years. Light, faster than which nothing physical can travel, takes years to traverse the distances between the nearest stars. Space vehicles take that long at the very least."[6] Well, how do we know? I mean how do we know how fast thought travels? The solution may lie in the parapsychological realm; the means of getting information I mean.

VALLÉE: There's another thing I'd like to bring up. It's from the "Report from Iron Mountain"—a part in this report that talks about political substitutes for war, and it says:

> This is where the space race proposals, in many ways so well suited as economic substitutes for war, fall far short. The most ambitious and unrealistic space project cannot of itself generate a believable external menace. It has been hotly argued that such a menace would offer the last best hope of peace by uniting mankind against the danger of destruction by creatures "from other planets or from outer space." Experiments have been proposed to test the credibility of an out-of-our-world threat. It is possible that a few of the more difficult-to-explain flying saucer incidents of recent years were, in fact, early experiments of this kind. If so, they could hardly have been judged encouraging. When anticipating the difficulties in making a need for a super giant space program credible for economic purposes, even where there are not ample precedents, extending it for political purposes to include features associated with science fiction would obviously be a more dubious undertaking.

And the reference they give for the argument of outer space menace as a best hope for peace is Robert Harris in "The Real Enemy," an unpublished doctoral dissertation made available to the study.

HYNEK: We can get that from the University of Michigan microfilms.

HASTINGS: If that's a genuine reference. But that's a common enough argument. What indication is there from sightings and observations which have given some hostile intentions? Incidentally, that never comes up; nobody ever seems hostile.

---

[6] Carl Sagan, *UFO's, the Extraterrestrial and Other Hypotheses*, CRSR463, Cornell University Center for Radiophysics and Space Research; Ithaca, New York, September 1971.

HYNEK: Well, actually, they do in some cases.

VALLÉE: People shoot at them—the number of incidents where people have shot at UFOs is significant.

HASTINGS: We do, but what do *they* do?

VALLÉE: Well, they stop cars, they paralyze witnesses. The way Charles Bowen[7] was putting it the last time I spoke with him in London, you have to assume there is something evil in an object that flies at low level above houses at night.

Are UFOs Hostile?

HASTINGS: Do they cause physical harm sometimes?

VALLÉE: Yes. A man named Floyd in the Everglades got too close to one, and a beam of light hit him between the eyes. He was blind for several days.

HYNEK: Very few permanent injuries, that we know of anyway.

VALLÉE: Except for the cases in South America, where some people allegedly developed nausea and strange symptoms and died a couple of days later.

HYNEK: One case that deserves mention here was brought to the attention of the Center through a medical doctor who works for the United Nations in Ethiopia. Let me read you his letter:

> I want to inform you about a strange natural phenomenon that happened in the village of Saladare, about fourteen kilometers away from Asmara, on the 7th of August 1970, at 11:30 A.M. This is how the eyewitnesses—local villagers—described it. After 11:00 A.M. they heard a noise coming from the direction of the nearby forest which sounded like an airplane flying low. The noise increased to an earsplitting strength and they saw a red glowing ball which swept through the village destroying houses in its path. It uprooted trees, burned grass, but didn't cause fires. When it left the village, it melted the asphalt on the road in an approximately two by seven meter area and broke the stone wall of the bridge to pieces. Then it travelled farther about 150 meters toward a hillside where it became stationary and then hovered above the ground for a few seconds and then started to travel back almost on the same route or rather parallel with it. On its way back it destroyed other houses and disappeared almost in the same direction it came from. The total distance it travelled in view was about three kilometers there and three kilometers back and the whole show lasted about ten minutes.

---

[7] The editor of the London-based *Flying Saucer Review*.

Some people claimed it had a tree-trunk shape and the inhabitants of the other nearby village said it flew over their village with an earsplitting noise. According to them it was ball-shaped with a tail behind it.

The excitement was so great that we went out three times to look at the village and I took about thirty photos and I am sending a few to you. It looks like a cannonball shot through the houses. *Fifty buildings got damaged, eight people injured, one little child died.* [Emphasis added.]

Some people think it was a meteor, but that wouldn't travel to and fro. It couldn't have been a tornado type of wind because it didn't carry away the tin roofs, they stayed in the same place all crumpled and melted and distorted. Up to now we have no idea what it was. The "Assis" (Addis Ababa) paper reported it in a few words as a thunderstorm. I am sending you the clipping. The local Italian paper reported it in five columns. Discarding the idea of wind or lightning, the weather was clear, bright, cloudless—so lightning is out. The village—same as Asmara—is up on 2,300 meters altitude, for example. Lightning sometimes travels horizontally, but as I mentioned the weather was clear. On the other hand, it had a source of heat. It melted the asphalt, metal objects, grass and bushes were burned, but without fire or flame. Its mechanical impact was tremendous. It went through the half-meter thick stone wall of the bridge and after turning back it still had energy to do damage....

This is one of the few documented cases where harm has been caused by something we must regard as a UFO. It was certainly flying, it was obviously an object, and it certainly was unidentified.

HASTINGS: But in general, more hostile actions have come from the people, as in the case of Mr. H.; humans have attacked them, shot at them!

VALLÉE: Yes. But if you believe Keyhoe, there were several incidents of encounters between jet fighters and UFOs, where the plane was destroyed. There is one particular case in Walesville, New York, on July 2, 1954, that fits this pattern.

HYNEK: That one isn't documented.

VALLÉE: Yes, it is documented! It was even mentioned in the *New York Times*. There was a picture in the *New York Times* the next day, of the town in flames with the main engine of the jet in the town square—four people were killed in the crash.

HYNEK: The Air Force, when I tracked that down, simply attributed it to mechanical malfunction.

VALLÉE: Quite true, it was a mechanical malfunction all right, but the malfunction was caused by a UFO. In this case in Rome, New York, July 2, 1954, there was a dense cloud cover.... On the radar they picked up an unknown object. They sent two jets up, one stayed in the clouds all the time, the other one was at a higher altitude looking for the object and broke through the clouds and saw the object. The two pilots said that they saw the thing, and it was coming close to them, and then they described something they called a "heat wave"; the temperature rose in the cockpit so that they couldn't operate the airplane anymore. They could hardly read the instruments, so they thought the plane was going to burst into flames. Well, they bailed out and landed safely and are still alive.

HYNEK: I wonder what our chances are of following up on a case like that. It would require a full-fledged operation with lots of money that could put a detective on the case. It would take him weeks perhaps to find these guys, but . . .

VALLÉE: Well, I had obtained their names and phone numbers, and I gave that to Condon in 1966. It was the first thing I gave to Condon, telling him: "You have the authority, you can pick up this phone and call those guys. Here are their phone numbers, here is a perfect case for physical evidence." He never did it.

HYNEK: Well, they may have moved since then, but we'll try it.

HASTINGS: Unless they've been visited by three men in dark business suits, like Mr. H.

VALLÉE: Anyway the plane kept going, just flying great until it crashed in the town of Walesville, killing four people. And nothing more was heard. There was a picture the next day in the *New York Times* of houses in flames, with the main engine of the jet in the square. You see, that to me was the test. If the University of Colorado really meant business, if Condon really wanted to know, he would have picked up that phone. All he had to do was dial two numbers.

HASTINGS: The Iron Mountain Report was a deliberate "put-on." The author wrote it to make evident a lot of assumptions that could very well be made, whether or not they were true. The book itself is not genuine, it's a scenario. It was deliberately designed as a social impact study to make the point that war is good for society in certain ways, and if we're not aware of that, we may unconsciously allow it to be reinforcing other negative patterns. That's the way I look at it. I think Robert Heilbruner was one of the authors.

VALLÉE: Who funded this?

HASTINGS: I don't know. What about UFOs expanding consciousness, in terms of social impact?

HYNEK: We could do a little blue-skying, as to what we think might happen. There is a certain disenchantment with science in general among people. This doesn't extend to astronomy. Astronomy has come out smelling like a rose because you can't blame pollution on it, you can't blame anything on it, and in the last decade or so, many exciting things have happened in astronomy that are capturing people's imaginations; quasars, pulsars, molecules in space! Black holes! Here again is a beautiful example; no astronomer in his right mind would have ever thought, ten years ago, of getting up and giving a paper on the possibility of complex molecules like formaldehyde and alcohol in space. (And then these blasted radio telescopes incredibly came up with over thirty complex molecules found in space.) There seemed to be no way in which complex molecules could have formed in the essential vacuum of outer space—and there was no indication they existed. The discovery of these molecules through radio astronomy came as a complete surprise. In so many new ways, astronomy continues to capture and expand people's imaginations, particularly the imaginations of young people. This again is another point for the existence of life in space because the raw materials are there!

I know that particularly the kids are intrigued by two series of things these days: From the questions we get at the observatory and in the high schools I talk to, and in freshman classes, the main question areas are the black holes, quasars, and pulsars, and UFOs, bunched together. That is what grabs them. And so, to that extent, both astronomy and UFOs are expanding imagination and consciousness.

6
Flying Saucers You Have Swallowed

Toward a Social Psychology of UFOs

An Air Force Hoax

The purpose of the experiment conducted by the Air Force over Clearwater, Florida, on July 4, 1964, was to investigate the accuracy of reports by individuals who witness aerial phenomena. The case offered the desired level of control, because no less than eighty witnesses were available. These people submitted letters in response to an Air Force request for information about six "mysterious" red lights seen over the sea. These red lights were, in fact, flares purposely dropped by an Air Force pilot. A statement by the pilot was released later. This statement detailed the phenomenon and afforded complete facts. This information in turn was used to evaluate the witnesses' reports.

Here is the pilot's statement:

> At 2005 [local time] on 4 July 1964, [name withheld] flew his Cessna 170-A aircraft from the Municipal Air Park in Clearwater, Florida. When over the Gulf of Mexico, approximately five miles west of Clearwater beach at 2030 hours, the pilot released five railroad flares attached to separate plastic parachutes. At an altitude of 10,000 feet, the flares were dropped in ten- to fifteen-second intervals, forming a line from south to north, parallel to the coast. With the aircraft travelling at cruising speed (120 mph) and the average time between drops being twelve and a half seconds, the total distance covered was two and two-tenths miles. This would space the flares roughly 2,000 feet apart. The

flares were of the common variety whose duration is ten minutes. Considering the twelve-and-half-second interval between drops, it can be said that the last flare burned out sixty seconds after the first. The duration of the sighting was therefore from ten to twelve minutes. The pilot proceeded to circle the flares until they went out at 7,500 feet. After losing 1,000 feet altitude, one flare dropped quickly and extinguished. To an observer standing five miles away, five degrees of arc would have been subtended by the lights as they fell from 10,000 to 7,500 feet over a period of ten to twelve minutes.

This being the case, then, thousands of people were in a position to observe the objects, clearly an uncommon and unexpected sight to them.

At that particular time there were two major layers of wind over the Gulf of Mexico, five miles west of Clearwater. The general flow was weak with an onshore breeze occurring only in the very lowest level. From two thousand feet to ten thousand feet, the wind was from the southeast at five knots and described as very weak. Even assuming a 50 percent error in calculations, the maximum distance the flares could have been blown toward the beach was less than two miles. This would place them over the gulf about three miles west of Clearwater beach at the time they extinguished.

The Analysis

Eighty witnesses responded to the request for information published by the Air Force in the local press. What did they report? How accurate were the descriptions?

First, let us consider the time of the observation: nearly one out of two persons responding on this point gave a correct time. Next, color: there was nearly 100 percent agreement that the lights were red. Of the aircraft lights, only 28 percent saw the white one and 19 percent saw the green one. There was a small tendency to confuse green and blue. Only three witnesses, however, saw all three colors.

Most witnesses stated they had seen six lights—perhaps counting the aircraft along with the flares. Sixty-nine percent of them gave the number as either five or six. Nearly 100 percent of the persons reporting on the formation of the lights described it accurately as a straight line. Only two persons estimated the separation between them. One gave ten feet, the other one hundred to two hundred feet (actually separation was about ten thousand feet).

The most inaccurate statements concerned the duration of the sighting (a large number reported twice the actual duration) and altitude. None of the witnesses reporting altitude was within 2,500 feet of the correct figure. The report comments, "The fact that only 11 percent of the witnesses made any attempt to estimate the altitude indicates the inability of observers to judge this detail."

One-fourth of the witnesses identified an airplane circling around stationary lights. Almost half of the observers reported a "weaving light." A parachute was seen by one person, and two stated that the lights floated "as if suspended by a parachute."

What is remarkable about these reports is not only their accuracy, but the fact that these witnesses had no idea of the nature of the phenomenon they were observing: Only two stated that the lights *appeared as flares,* and two stated that they were possibly caused by flares. Ninety-five percent of the observers could not identify the lights as railroad flares from a distance of five to eight miles.

The movement of the flares (that were almost stationary and extinguished at high altitude) led to several inaccuracies. Three of the witnesses stated that the lights fell into the sea, while four saw them "move off into the distance." Two observers insisted that they fell with considerable speed.

In trying to estimate the accuracy of reports, the Air Force set up a rating scheme, taking into account the number, color, and movement of all the lights, including those on board the aircraft. To have a "Very Accurate" report, the witness

> must specify that there were five large red lights over the gulf. He must either state that an aircraft was observed orbiting the lights or describe very accurately the motions of a weaving light which should further include an observation to the effect that the weaving light was similar to that of an aircraft or helicopter. His estimated time and duration must be no more than ten minutes off the known times. This description of the motions of all lights must coincide with the known movements, and any other related detail must be accurate to a high degree.

Then the investigators defined as "Accurate" any report that did not violate more than one condition essential to a VA report, and did not contain misleading statements. A "Fair" report was one that neither added to nor detracted from the sighting of five stationary lights. Finally, an "Inaccurate" report was either one of poor quality or one that contained some statement of gross contradiction or exaggeration.

With these definitions, 12 percent of the witnesses were considered "Very Accurate" and another 37 percent were rated "Accurate." Twenty percent of the males were VA while only 8 percent of the women were in that category. Both male and female observers were 35 percent accurate, the women having about six times more inaccurate reports.

The Clearwater witnesses stated on several occasions that they would not have remembered the observation if they had not seen the article requesting information. They felt "it was their duty to comply with the Air Force's request for information." Given the fact that these people never knew what they were observing, how many of them let their imaginations take over? The answer is, extremely few. Apparently, there was only one reference to the idea of a threat by extraterrestrial visitors. Even in this case, the report is such that the true facts can be partially reconstructed from the information it gives. The letter said:

> I kept saying, watch that plane! Mrs. [name withheld] insisted they were balloons. We were watching fascinated, then the plane went to number one circle making a pass quite near it. I heard a loud noise. It disintegrated. Then the plane circled to No. 5 as marked on the enclosed paper, back to No. 2 then No. 4 then No. 3 and last No. 4. Each one disintegrating as the plane turned from its circling and went nearer the big red light.
>
> This seemed so fantastic. The lady who was standing near me exclaimed at the first sight of them, "Could it be the Martians are comming [sic]?"

It is just as important to note what the witnesses did *not* report as what they did. They did not report creatures, abduction, telepathic messages, luminous craft with portholes and antennas ... in short, by and large, they reported flares.

Did the Star Capella Land on the Runway?

This is a fine experiment, but many factors were left aside when the investigation was completed. It certainly is possible to judge the accuracy of the observations, but we really know little about the degree to which they are affected by interpretation and the meaningfulness of such measures of "reliability" as the witness's background and community status.

There have been articles by sociologists on this very point; they have explained—without quoting much direct evidence, it is true—that the people who reported such unusual phenomena as "flying saucers" were individuals craving recognition and status. Status inconsistencies, however, do not come close to explaining UFOs, and the second case we would like to describe will clarify some of this. It really complements the Clearwater experiment. In the next series of sightings, we have an opportunity to see how society reacts to the unusual on the basis of witness accounts that are reliable to a very large extent.

Our first contact with the case was in 1967, when we were going through volumes of Air Force archives containing reports of unidentified objects. We came across a series of observations by people in several northern states (Wisconsin, Minnesota, Illinois) describing a very peculiar object in the northern sky. The light was moving, said the reports, and at times it dashed towards the observers, scaring them seriously. The Air Force had "explained" these reports with a single interpretation: the star Capella.

Capella is a bright northern star that will remain close to the horizon for hours for observers situated in the northern U.S. and Canada. When it is close to the horizon, it scintillates and sometimes gives the illusion of sending signals of various colors while moving suddenly. However, it would be very strange for many observers in several states to send reports describing misinterpretations of the same star at the same time on a particular evening! We tried to get the Air Force to reopen the case, but this was hopeless. Finally, we decided to study it on our own.

The date was August 16, 1966. The reports that started coming in the next day (appearing in newspapers and military teletype messages) described a large ball of light that came close to several airplanes, flying alongside them for durations of two to thirty minutes, and then departing with changes of shape. The first reports came from Truax Field in Madison, Wisconsin. Later descriptions were given by people on the ground near Minneapolis, Rochester, Duluth, Hibbing, and other northern communities. One report stated that pilots at a major airport had seen the mysterious object land on a runway and then take off again!

Mr. William T. Powers, the chief systems engineer at Dearborn Observatory, took an interest in the case and proceeded to track down all the reports from pilots, whom he interviewed by telephone. All the other names mentioned in newspaper stories were investigated, and finally half a dozen papers in Minnesota and northern Wisconsin were asked to run an announcement requesting reports from anyone who had noticed anything strange the night of August 16th.

Powers obtained thirty new reports involving seventy-four people located over four states—from Illinois to North Dakota. They certainly had not all been observing Capella! At this point of our investigation, we knew the Air Force was wrong, but we still did not know what the people had seen. When the reports were put together, the following picture emerged:

> At about 10:45 P.M. (CST) on August 16, 1966, a small bright spot of light appeared low in the northern sky. It rapidly expanded, eventually became larger than the full moon and somewhat brighter. As it expanded, it changed colors from a reddish tint, through swirling blues and greens, to a final shade of silvery white, which then faded. The object retained a circular shape except for a brief period near the beginning, when it resembled a rectangle, and near the end, when it elongated and disappeared, leaving a slanted streak of smoke or light lingering behind. Its direction changed, if at all, only slightly and slowly. A few minutes after the end of the display, the same thing *happened again*, at a slightly higher altitude.

When we showed this to the Air Force, they told us we were wasting our time, but Powers continued his investigation by drawing lines of sight on a map; it became clear that the object had been far to the north of all the observers—in fact, it must have been outside the borders of the United States, in the Canadian wilderness beyond Minnesota! The most likely point was situated two hundred miles into Canada, and the object must have been extremely large and luminous to be compared to a full moon in Illinois! Also, he observed that the witness farthest south, who was flying over Illinois, had seen the object a few degrees above the horizon. It must, then, have been several hundred miles up in space. If the stories were to be believed, a great explosion must have occurred several hundred miles over east central Canada, at about 10:45 (CST) on August 16th.

Impossible? We found from our colleagues in Canada that a team of scientists had launched, on that particular night, two successive rockets from Fort Churchill, which is located on Hudson Bay. These rockets rose 250 miles and each released a cloud of barium. Such a cloud expands rapidly into the near-vacuum and fluoresces brightly from bombardment by high-energy particles. The timing, location, and description of the occurrence matched very precisely what was derived from the reports. Commenting on these descriptions Powers wrote:

> A case like this could reveal a wealth of information about how to interpret UFO reports. Every witness was awestruck by the event; none had seen anything like it before; none had the least idea what he was looking at. The object could have seemed no stranger if it had been a flying disk with portholes. It was a bona fide UFO and an astonishing one at that. We can, therefore, look at the reports as we would treat a report of a classical saucer, for the reactions of the witnesses were much the same as those of witnesses reporting unknown and equally striking events.

It was interesting to compare the reports given by technically qualified observers with those of unqualified people. The pilots reacted coolly even when they were startled. All but one of them interpreted the rapid expansion of the cloud as a rapid approach, and several took evasive action to avoid collision. Only a third of the non-pilots interpreted the expansion as an approach. Powers's report goes on:

> The data do not give very strong support to the notion that adults are better observers than children, or that responsible people will make better reports than those who have ordinary occupations. *One learns that a clear-cut distinction between "reliable" and "unreliable" cannot be made so easily.*

One witness coming home from an evening in town, who panicked along with his three passengers and drove fifty miles per hour backwards when he thought the object was about to attack the car, nevertheless *gave a creditable description* of the colors and shapes, and his report of the direction was accurate. *A young teenager gave a clear, accurate, factual report, concluding that he may have been observing "some kind of explosion."*

Two adults in responsible jobs, who described themselves as "scientifically trained," were off an hour in time, off forty-five degrees in direction, and since they rejected the notion of UFOs *concluded that they might have seen "lights reflecting off a fog bank!"*

An FAA controller saw the object and decided it was a reflection of the moon in the window of the control tower, despite the fact that control tower windows are slanted to prevent such reflections, and there was no moon in the sky!

A farmer gave a completely accurate report, including angles with error estimates, accurate times, and a precise description of the colors and shapes. He happened to be an amateur astronomer.

A few months later, some researchers on aerial phenomena heard of the case and got very excited. One of them went on to write a sensational article about the "flying saucer flap" of August 16. We tried to make him understand that the "flap" was all explained, and that he would better use his time, and his considerable talent, studying other mysterious reports. But he wanted to continue to believe, just as the Air Force wanted to go on with the Capella explanation. His article was published in *Fate,* and is still quoted today as evidence that outer-space visitors, once in a while, pick a particular region and cause hundreds of sightings.

The Caliber of the Witnesses

The fact that so many sightings of reported UFOs turn out to have natural explanations certainly helps explain the lack of interest displayed by many scientists. But given these sociological effects, what can we expect in the way of a breakthrough? This was one of our discussion topics:

HYNEK: I always kept hoping that a really crucial case would show up and convince the Air Force. That's one thing that kept me from saying anything for a long while. I was sort of hoping, maybe next week a really good case is going to come along that will really tear things open. That never came along, but what did tear things open, for me, was just the accumulated weight of evidence, more and more and more. When I turned in my first report to Project Sign, we had only 237 cases, four-fifths of which were more or less explainable. That meant we had altogether no more than fifty cases that were suggestive of something really going on. And then I thought, well, maybe the witnesses in those fifty cases were untrained people. . . .

The caliber of the witnesses, we've touched on it a little bit, but one thing we should dispel is the notion that UFO witnesses are a special breed of people. They are not. The UFO sighters represent a fair cross-section of the country. There is nothing pixieish about them—they're not all cab drivers; they're not all housewives. There is a good age spread. The majority of the witnesses are between nineteen and fifty-nine. In other words, the majority are mature people. Is there anything more we can say about the witnesses themselves?

VALLÉE: In the study Ballester-Olmos and I did of the Spanish wave, we had an opportunity to do a sociological study of witnesses in landing cases. Taking Spain as a case book, we made some observations about the composition of the groups. Most of the cases were reported by families, a man and his family. As to what they were doing at the time, most were driving on a familiar road—it turned out that most people were engaged in their usual occupations, in their normal surroundings. Coming back from work on the road they used every evening, going to pick up their kids, etc.

HYNEK: There was nothing that gave them any warning in advance—no special set of circumstances.

VALLÉE: They were not looking for it, they were not alcoholics or drug users. Now, what about drugs? What about UFOs seen by people under drugs?

HASTINGS: Never heard of any sighting reported where drugs were involved.

HYNEK: That shows how careful we have to be. You said something earlier, Jacques, about the gathering of UFO facts being a public relations job. I suppose we have to behave more like social scientists than like physical scientists. Take physics, for example. New data are far removed from the public; one or two investigators, after months of painstaking labor, using sophisticated equipment, come up with one tiny new fact—a new subatomic particle, for example—about which the public couldn't care less. It doesn't touch their lives. Yet that fact is immediately accepted at the highest levels of science because it fits the scientific belief structure.

What a difference with UFO facts! These come from the public and do touch their lives. Now it's the scientists who couldn't care less. Here human beings are the instruments through which facts are obtained and not through linear accelerators or mighty telescopes. UFO fact gathering is much more akin to the manner in which sociologists gather facts; one has to go to the people. But one has to be very careful about what people one goes to. There have been, and are, people like Adamski and many other highly nonobjective spawners of UFO tales who thoroughly muddy the waters.

VALLÉE: It seems to me people used to suspect the witnesses of being drunk. Now they suspect them of being stoned. That's a sign of the times! I have heard of cases like that—in California, it's impossible not to hear of cases like that, but they are not at all like the UFO reports we know; although sometimes the witness says, "I had taken acid the night before, but I was okay when I saw it." I have heard the case of a fellow who was smoking marijuana at a party. Suddenly, he appeared to leave his body, went into space, and found himself in this flying saucer where he met a character wearing an old-time sort of suit. He was about four feet tall, with a long white beard, white hair, and nicotine stains on his beard. He was like a little leprechaun. He said, "Hi, I'm God . . . as a matter of fact, this whole spaceship is God!" And the guy said, "Groovy, can I stay?" But God said, "No, you've got to go back!" He found himself going backwards through space. He could see the stars and there he was! His friends around him saying, "What happened? You passed out." So I've got cases like that but not exactly of the highest level of reliability! [Laughter.]

HASTINGS: High strangeness, low reliability!

VALLÉE: But here you have direct contact with God, direct data. God had nicotine spots on his beard!

The Life and Lies of George Adamski

HASTINGS: And what about George Adamski? Can we review your notes on that?

VALLÉE: Relatively few people know the whole life story of Adamski. He was born in Poland on April 17, 1891. He was two years old when his parents settled in Dunkirk, New York, where he grew up. In 1913, he enlisted in the Army. He served in the 13th Cavalry on the Mexican border, receiving an honorable discharge in 1919. Two years earlier, he had married Mary A. Shimersky, and he was nearly forty when he settled in Laguna Beach, California, and devoted full time to "teaching the universal laws." At that time, he founded a mystical cult called the Royal Order of Tibet. In 1940, Adamski and a few of his students moved to Valley Center, where they established a small farming project. In 1944, the group moved to the southern slopes of Mount Palomar.

I followed Adamski's public career after the publication of his book *Flying Saucers Have Landed*. At the time, most people in Europe thought he really was a professional scientist in America. I recall an article in France's authoritative journal, *Le Figaro*, quoting him as "Professor George Adamski, of Mount Palomar Observatory."

Adamski did have an observatory on Mount Palomar, you see. But of course it was not *the* one! There was nothing to prevent him from calling himself "Professor," and no law could stop him from erecting a small dome with a toy telescope inside.

The rest is a tribute to the journalists' eagerness to find someone in authority who would be willing to talk about flying saucers. George was more than willing!

HASTINGS: How did he become a contactee?

HYNEK: His involvement with his space friends began, as far as we know, on November 20, 1952. According to a man who published an investigation of Adamski,[1] he had tape recorded a "mystical" transmission that instructed him to go into the desert, which he did, accompanied by Alice Wells, owner of the Paloma Gardens Cafe, and his private secretary. Mr. and Mrs. George Williamson of Prescott, Arizona, and Mr. and Mrs. Al Bailey of Winslow, Arizona, came along too. Adamski instructed the others to stay near the car while he went up a small canyon. When he returned, he claimed to have seen a landed saucer and to have conversed telepathically with its occupant, a "man from Venus." He took seven photographs that turned out blanks. The six members of the party signed an affidavit, swearing they were witnesses to the event. Traces left by the shoes of the Venusian have been carefully preserved!

VALLÉE: But a few years later,[2] Mr. Bailey, who was a railroad worker, stated that he had seen neither a flying saucer nor a stranger near Adamski, and that none of the others could have seen more than he did from their location.

HYNEK: Anyway, the same article says that the next month (on December 13, 1952) George's Venusian friend flew low over Adamski's property at 9:30 A.M., and Adamski thus obtained several telescopic photographs of the saucer. Sergeant Jerrold E. Baker is said to have independently photographed the object with a Kodak Brownie camera. Now the plot thickens, because the photographs taken by Adamski could not have been obtained with the equipment stated.[3] Furthermore, on June 29, 1954, Mr. Jerrold Baker, who was a member of Adamski's household at the time of the desert expedition and of the alleged flyovers, wrote and swore that, "I did not take the alleged photograph accredited to me."

In November, 1953, Adamski had written to Baker: "Now, you know that the picture connected with your name is in the book, too—the one taken by the well

---

[1] See articles by James Moseley, Irma Baker, and Lonzo Dove in *Saucer News*, "Special Adamski Expose Issue," October 1957.

[2] In a letter dated June 1, 1954, to Jerrold Baker, and in interviews with James Moseley.

[3] In the words of Arthur C. Clarke (*Journal of the British Interplanetary Society*, March 1954): "Many people, including, we suspect, Mr. Adamski, do not realize that a large object seen through a telescope bringing it to within twenty feet looks quite different from an object itself twenty feet away."

with the Brownie. And with people knowing that you are interested in flying saucers as you have been, and buying the book as they are ... you could do yourself a lot of good. For you have plenty of knowledge about these things, whereby you could give lectures in the evenings. There is a demand for this! You could support yourself by the picture in the book with your name."

VALLÉE: What I like most about Adamski is the description of his friends from outer space. One evening in December 1952, he had gone to Los Angeles, "drawn by a strange mental impulse," and he was approached by two men—a Martian and a Saturnian—who took him to the desert in a black, four-door Pontiac sedan. They went to a glowing saucer where they met Adamski's Venusian buddy, who said: "As we were coming down, a small part of this little ship broke, so I have been making a new one while waiting for you to arrive." So saying, he emptied the contents of a small crucible onto the sand. They took off and visited a mothership in space.

Then in April of the following year, George "again felt a sudden urge to go to the city." Once in Los Angeles, at 7:15 P.M., he met his Martian friend Firkon and took him to a cafe, where Firkon ordered a peanut butter sandwich, on whole wheat, black coffee, and a piece of apple pie. Later they drove to the hills in the usual black Pontiac and boarded a "scout from Saturn": that took them to a Saturnian mothership from which Adamski had the privilege to observe the moon through a powerful telescope: "Many of the craters are actually large valleys, surrounded with rugged mountains." He was told that there was still plenty of water on the other side.

And later, in August 1954, Adamski again boarded a spaceship where two girls from Venus gave him some fruit juice. They were dressed in pilot uniforms, hence, "I felt sure this meant a trip into space." The other girls, however, wore lovely gowns. They flew around the moon, where they saw lakes and rivers and "a number of communities of varying sizes." Finally, there was a farewell banquet for Adamski.

HASTINGS: Is he still alive?

VALLÉE: No, he died in 1965, but he has already communicated from the other side through several mediums!

Dr. Hynek Meets "Professor" Adamski

HYNEK: Did I ever tell you about my meeting with Adamski?

HASTINGS: No.

HYNEK: What happened was that several astronomers and I were going up to Mount Palomar, the real Palomar, and we'd all heard about Adamski, so we stopped at the little cafe where he peddled his photos. We had agreed that within his earshot we would start talking about some special filters we were going to use in the two hundred-inch that could prove there is life on Mars. And it was hard to keep serious, but we started talking... Oh! were Adamski's ears flapping in the breeze! He came over and introduced himself, and I'm sure that he told people afterwards that he had it straight from the top astronomers that there was life on Mars! But then, as though I didn't know him, I started asking some questions which quickly led into the UFO business. All he wanted to do was to sell me some pictures! He showed me his pictures of UFOs against the moon, and I said, "Well, what emulsion did you use, what speed, what camera?" and so forth. No, none of that. All he wanted to do was sell me photos. He said he'd made a trip to the moon, and I was on the verge of saying, "Well, damn it, I'm an astronomer and we just *know* that there's no vegetation on the other side of the moon!" He reported that there were people and there was vegetation and so forth. Well, that at least is reporting something that could be tested. It came out negative, of course, but most of the contactees don't give you any testable information—it's platitudes. Ban the bomb! Be nice, we're not going to hurt you, we're coming to help you. "Platitudes in stained glass attitudes!"[4]

Is a Martian Buried in Texas?

One of the earliest flying-saucer hoaxes on record was perpetrated in Texas in April of 1897. There was in progress at the time a fantastic wave of rumors about mysterious "airships," similar to our flying saucers, that flew in the skies of California[5] and the Midwest. This series of sightings (we have several hundred unsolved cases on record) was the occasion of a number of jokes, and in Chicago it gave rise to the first faked picture of an Unidentified Flying Object.

It was on April 17, 1897, at 6 A.M. that a strange object flew slowly over the public place of Aurora, Texas. The intruder hit a windmill and exploded, says the Dallas *Morning News* of April 19, 1897, "destroying Judge Proctor's flower garden. The badly disfigured remains of the pilot were found, and he is clearly not an inhabitant of the Earth, but probably comes from Mars, according to Mr. Weems, an authority on astronomy. Papers found on his person were undecipherable. Mr. E. E. Haydon reports that the ship is built of an unknown metal, resembling somewhat a mixture of aluminum and silver."

---

[4] See *UFO Controversy in America, 1897–1973* by David Jacobs, Indiana University Press, 1975, for a detailed history of the contactee aspect.

[5] For the details of the California sightings, see "The UFO Wave of 1896" by Loren Gross. Privately printed. Fremont, California, 1974.

This case was researched by Donald Hanlon and the authors of the present book in 1966, and an investigator went to the scene—admittedly a long time after the crime! Here is what he told us:[6]

I drove to Aurora and stopped at the only service station there. The proprietor of the service station is named Oates, and the station and his house are on what was Judge Proctor's place. It seems as if Aurora was once the largest town in Wise County and was also the county seat, and Judge Proctor was a Justice of the Peace. Mr. Oates would neither deny nor confirm the authenticity of the story, but he told me to contact a Mr. Oscar Lowry a few miles down the road in the town of Newark, Texas, which, incidentally, is another thriving metropolis of slightly over three hundred inhabitants.

I found Mr. Lowry just where I was told I would, "down the road apiece by the schoolhouse." When I told him what I wanted, he asked me to sit down on a bale of hay (I had found him in his barn) and he would tell me the story. He was about eleven years old when it had happened, and he also told me that no less than twenty others had been there before me.

Mr. Lowry said that Aurora was a busy little town until the railroads put down their new tracks and neglected to include Aurora in their plans. As a result, the town began to diminish as people moved to be near the railroad. E. E. Haydon was a cotton buyer and writer who lived in Aurora and wanted to do something to help keep people in town and to make it a tourist attraction. He got the idea, I suppose, from the actual sightings he had read about and made up his story. The T. J. Weems that was supposed to have been a U.S. Signal Service Officer was actually the town blacksmith and, according to Mr. Lowry, the Proctor place never had a windmill on it. To further substantiate the hoax, the cemetery is a Masonic cemetery and a charge is kept on who is buried there. There are no unaccounted for graves. Mr. Lowry said that Mr. Haydon later told others about his story and many went on letting people believe it.

Such were the hoaxes in the good old days. Nobody really gained anything from it. The present century is more practical. Some of our modern "contactees" have made a sizable profit, not only from the publication of their stories, but from subsequent lecture tours. They also have founded "research institutions" that dispense courses on a variety of subjects, from "Soul Growth" to "Cosmic Wisdom." These activities, often barely legal, are the basis of a thriving business, and not just in California.

---

[6] See Hanlon & Vallée, "Airships over Texas," *FSR* 13, no. 1 (Jan.–Feb. 1967).

Planned Invasion Delayed

The time was bound to come when some enterprising pioneer would capitalize on the fears and hopes of the public, and would attempt to hoax his way to fortune.

In 1959, a man named Karl Mekis, fifty-two, self-styled *Venus Security Commissar on Earth*, was found guilty on seventeen counts of fraud and swindling and sentenced to five years in jail by an Austrian court. For six years, he had exploited public credulity, sold thousands of "survival kits," as well as jobs in the post-invasion "World Republic of Venus," and had accumulated a fortune in excess of $300,000.

His history was simple: He had met the Space People who had come to Earth from Venus in their flying saucers. They had revealed to him their plans to invade the planet and enslave all the Earth people, and they had put him in charge of security in their new government. A member of Adolf Hitler's SS guard during World War II, Mekis was not ignorant of terror techniques. But he did not have much success in life until he decided to move to South America. On the boat, he met another confidence man, Frank Weber-Richter. Together they established the plans for the Venus invasion. Soon after their arrival in Santiago, they set up their headquarters and started their campaign.

The first announcements of the coming invasion appeared in European science-fiction magazines: their advertisements gave such warnings as these:

> Only a handful of Earth people will be picked to rule with the Venus masters after the invasion. And you can be one of them—if you act now.

Other ads were more specific.

> World Republic of Venus. Chauffeur wanted to serve top official of Venus after invasion of Earth.

Some advertisements even offered marriage to Earth women with Venus men, to build a new "master race." The invaders were white and were said to "speak in a very clear, intelligent tongue." Thousands of people believed Mekis. A twenty-two-year-old Austrian secretary worked for him without pay because he promised her a Venus man after the invasion. "I thought everything he said was true because he went into such detail about his talks with the Venus people," she said. Other victims sent him their life savings in return for identification cards giving them special privileges after Day X, the day the Earth would fall into the hands of the Venusians. Mekis and his friend soon were making thousands of dollars per week. But the growing interest of the Chilean police in the whole operation motivated Mekis and his partner to move back to Europe.

They settled in Rome, but carefully avoided doing business there. They explained to their after-invasion staff that the Venus planners had ordered them to move closer to their European followers. Money continued to flow. Later, a variety of reasons were invoked to justify postponement of the invasion, including the sudden death of the Venus commander in chief! All these announcements called for "more volunteers and more funds." Mekis was arrested when he left Italy for a vacation in Austria. At his trial, a forty-year-old Bavarian tavern owner produced this telegram from Mekis:

PLANNED INVASION DELAYED. NEED FINANCIAL SUPPORT. PLEASE SEND ALL YOU CAN.

The victim further testified:

I was the newly appointed Adviser for Venus Economic Affairs, so I thought I had to do something. I took my last $1,100 out of the bank and sent the sum to him.

Other witnesses presented their passports, showing they were official citizens of Venus. Mekis was sent to jail. Weber-Richter escaped. But the "contactee business" is still booming. In California and elsewhere, "research centers" have sprung up, devoted to the repetition, under a legal front, of the process developed by Mekis—capitalizing on the fear of the unknown, on the technological-cultural gap. And one has only to read the bulletins of certain UFO clubs and see the items they advertise (up to $7.50 for a thirty-minute tape of' "space messages" such as "Universal Vibration" by Esu and Monda, "To Men of Earth" by Voltra, and "Elementary Magnetics" by Bellarian!) to realize how fear and anguish are widespread in this Age of Space, and how many people have already jumped beyond the edge of reality!

7
Reminiscences

The Scientific Debate about UFOs

The Invisible College

The state of the art in UFO studies has undergone tremendous changes in the period from 1952 to the present. The authors were part of that evolution, and the following discussion was initiated to bring out their candid appraisal.

VALLÉE: Do you remember the early days of the "Invisible College" and what it was like to work on UFOs when the subject was underground?

HYNEK: And we met at your apartment on Bryn Mawr Avenue, in the mid-sixties?

VALLÉE: And the attitude of all our scientific colleagues . . .

HYNEK: Well, they didn't know too much about any meeting we had, because those were certainly "underground." What little they knew about it, they simply regarded as an aberration, as an indulgence, "If they want to do that crazy stuff well . . . they could be collecting ancient guns or something. . . ."

VALLÉE: What about the people like Menzel and so on who were very keenly interested in the subject?

HYNEK: Menzel all through has maintained a rather interesting friendship with me. I never had any acrid letters from him; in fact I just have a recent one from him. He's just now gone off to Costa Rica. He is very much interested in what I found in the Father Gill case in New Guinea. Of course, he feels that Father Gill is myopic or something, but he'll be very surprised when he reads that I found some of the original witnesses.

VALLÉE: Isn't it interesting that most of our colleagues pick one case here and one case there, either good or bad, and that is everything they know about the subject. They will ignore 15,000 cases because of that one particular case they know. Their landlady said she saw something and they went outside and it was Venus, so that explains everything as far as they are concerned! Suddenly they think they don't have to investigate further because everything is solved.

HASTINGS: My perception is that Menzel feels obligated to deny any reality to UFOs, but he may expect to be proven wrong.

VALLÉE: Do you think it's necessarily true that young people are more willing to accept this than established scientists?

HYNEK: Yes, very definitely.

VALLÉE: How come more young scientists are not studying this?

HYNEK: I think many would like to, but there is no promotion in it, there is no journal that will publish it. Well, take H*** for instance. He's quite interested, but he couldn't publish a paper on it in the *Astronomical Journal* or something of that sort. The young are more open-minded.

VALLÉE: How was it when you worked with Ruppelt?[1]

HYNEK: That was in my debunking days, and he regarded me with a certain amount of suspicion. I was just one of these professors who were coming around. He never really took me into his confidence, except occasionally we would have a beer together, and he'd talk a bit, but it was somewhat of an ego trip for him. He was constantly in demand, briefing this general and that general. If I had to give an impression of him, it would be that he was sort of a weathervane. He was extremely puzzled, and one day he was in one direction and the next day his weathervane had shifted to the other. He was trying to do the best job he could to debunk it and yet he had this weird perception ... something was going on that was beyond him. He wouldn't communicate with professors, and he didn't care much for Menzel because Menzel came around and simply said he had it all explained and Ruppelt was too smart to buy it.

VALLÉE: I remember going to Wright-Patterson with you once and the feeling about this office was that they were very much doing a routine public relations type of job. It might have been very different if there had been full-time scientists there who were interested in the work.

---

[1] The late Captain Edward Ruppelt, the first director of Project Blue Book, which was initiated in 1952.

HYNEK: That certainly is true. As I said before, not even a major general wants to be laughed at by a Harvard professor. They were taking their lead from the scientists, and if the scientists at that time had said, "This is serious, we should look at it," I think the Pentagon might very well have had a different attitude. But the Air Force had their own scientific advisory board, people like Whipple and Menzel were on it, and they said there was nothing to it. So the Pentagon took its lead from them and, of course, I think it sort of went . . . I think they were immensely relieved; I never thought of it this way before, but the Pentagon and the Wright Field people were relieved to have the scientists say that because it took the monkey off their backs. The top scientific brass said there was nothing to it and the military brass themselves didn't understand it; it was much too difficult for them, much too complex. Military problems in general are much more straightforward . . . they were happy to have that attitude expressed by the scientists.

VALLÉE: What determined the cases you could go out and study in the field?

HYNEK: That depended entirely on who was in charge at the time. I was a consultant, and I was very rarely given my head, so to speak; I could suggest that a certain case ought to be looked into, and it might or might not be. I never went out on a case with Ruppelt. I was just sort of brought in once in a while. Then there was the hiatus . . . wait a second, I'm wrong on that . . . I was, of course, the first consultant to *Sign* and that changed into *Grudge*. I was then off for about two years, and then when the big flap of '52 came, Battelle Institute had the contract for a study, part of which is still classified . . . I was sent around the country . . . they wanted to find out discreetly what astronomers felt about UFOs.

I remember there was a meeting at that time in Victoria, and I took the Empire Builder and went there. Whenever possible I brought up the subject in cocktail gatherings and in meetings. I travelled down the West Coast, visiting Lick Observatory and Mount Wilson and so forth, and I tried my best cloak and dagger, completely unobtrusive manner. . . . I think I questioned something like fifty astronomers, and asked what their opinions were. They varied; some of them were quite open-minded; I had a code designation for each astronomer, and I submitted that report to Wright Field.

About late '52 is when Blue Book actually came into being, and I became scientific consultant; I remember when Ruppelt would come to the Faculty Club at Ohio State. He visited the observatory a few times, and I visited Wright Field since we were just sixty miles apart.

One day he brought with him Captain Hardin and introduced me, saying that Hardin would be the next guy to run Blue Book. Hardin was a very soft, easy-going guy; by that time it was just after the Robertson panel, in the spring of '53, I believe. Hardin was strictly a career man who looked only for one thing, and that was the day he could retire. He wanted to be a broker, and he spent much time reading stock market data. It was quite clear that he was a totally different personality from Ruppelt. And, of course, the atmosphere had changed. There was still some sort of puzzlement in the wind, but after the Robertson panel, which was the big turning point, the orders had now come down from on high that this subject was to be debunked. It was supposed to be played down. It was simply, you find and fit whatever natural explanation you can to any and every report that comes in. It was no longer a question of finding out if there was something strange going on. What *appears* to be strange *must* come from some natural thing. To Blue Book this became a finger exercise in fitting square pegs into round holes or vice versa. They force-fitted a solution, despite the fact that they were honest enough to regard some cases as unidentified or unknown. There were some square pegs that just couldn't be fitted into round holes!

Don't Rock the Boat!

VALLÉE: What could have changed that? If one of the scientists had spoken otherwise, say Luis Alvarez for instance...

HYNEK: Well, that could have changed it... I don't know whether a single one would have, but if Robertson himself, or some of those... Thornton Page was a junior member essentially, he was the youngest member on the panel.[2] If they had changed their minds, something positive could have been accomplished. Hardin did a simply marvelous job of not rocking the boat. He was an Air Force man but he didn't have to fly, hated flying! The only places he went, he went by train, and we had some long journeys... Bismarck, North Dakota, and Colorado Springs... he simply hated flying.

VALLÉE: That's amazing. Because I think to most people in other countries, certainly in France, for instance, the image we had of the U.S. Air Force investigation was really something! We had this concept of a big project staffed by a dozen officers, with you as one of many consultants who could instantly get anywhere... jets waiting at the door... as soon as there was something on the teletype, the pilot would be there and you would be in the next seat and the plane would take off immediately and you would be flown to the site with all the equipment to start investigating.

HYNEK: Well, it was a fact that there was no attempt to be scientific....

HASTINGS: Instead you got someone reading the *Wall Street Journal* ... and you took the train!

HYNEK: Yes, and we took the train and Hardin himself couldn't hold with science. He was an affable fellow, very nice, but not the man for the job.

---

[2] Dr. Thornton Page was employed by NASA in the Manned Spaceflight Program in Houston until 1976.

VALLÉE: Now, when a local intelligence officer at an Air Force base, quietly sleeping at his desk in North Dakota somewhere, suddenly gets a report that there is a large object approaching from the north, he must do something about it, mustn't he? He can't say, oh, well, it's just probably one of those silly UFOs ... in a way the security of the country depends on what he does next. He's trained to go out and investigate these things. ...

HYNEK: Well, the view was that if the case could be solved at the local level, it would not go to Wright Field.

VALLÉE: But most of those are *not* solved. ...

HYNEK: Yes, and this is why Wright Field would get sometimes several hundred a month, but more generally twenty or thirty a month. Now, not being solved at the local level has two interpretations. Some really couldn't be solved at the local level, the really "wild" ones—ones that just couldn't be explained, no matter how hard they tried, as misidentifications of natural things. In other cases, the personnel at the air bases were just passing the buck. The local officers just didn't want to be responsible and didn't even attempt an explanation. They would say, in effect, no point getting involved. So they'd pass it on to Wright Field.

VALLÉE: What would have happened if around 1955 or so you had taken a handful of the best reports and had approached some of the more open-minded scientists in the country? Some of those who were respected in their field, maybe other than astronomers or physicists. You could have said, "Look, I think we have a problem here. The Air Force is receiving all these reports and, frankly, I'm getting concerned. We should do something." What kind of reception would you have had? You may have tried it. ...

HYNEK: No, I didn't try it ... I didn't seriously consider that because I felt rather helpless about it. I was still very puzzled myself. It's true it's probably my fault because I vacillated ... there were times when I would get a simply marvelous and tremendous case and then a whole run of terribly bad cases, like a gambler who suddenly has a run of bad cards; case after case after case of nothing, and I said, "Well, it looks like it is full of nonsense after all," and I would go down into a UFO depression phase, and a few weeks later a hot case would come in and the barometer would go zooming up again!

Of course, the late James McDonald[3] thought there were things I should have done; Jim McDonald certainly berated me tremendously. He said, "You were the scientist on the job, you should have called this to their attention." I think if I had

---

[3] A University of Arizona physicist, Dr. James McDonald wrote several articles on the reality of UFOs.

been twenty years older, a full professor, and a member of the National Academy, I might have risked it, and I might have been listened to. But at that time, I was either an assistant or an associate professor at Ohio State, and Ohio State is not Harvard, nor the University of Chicago; frankly, I was much more concerned with my own career at the time.

I knew if I really came on waving my arms I would have been declared a nut and my services would no longer have been required. I don't think I would have done any more good. Certainly any future effectiveness would have been shot down. I never would have been asked by Whipple to take charge of the satellite tracking program; I never would have gotten to Northwestern, because the reason I became the director at Northwestern was because I had a reputation in satellite tracking, with Sputnik and so forth. My temperament is to play the waiting game; I always have, maybe it's my Czech background.

VALLÉE: On the other hand, in those instances when people who had access to similar data went to the scientists with it, the results weren't very encouraging, were they? To some extent NICAP tried to do that.

HYNEK: NICAP had Admiral Hillenkotter on their board of directors; he was the former head of the CIA and he couldn't do anything. When the former head of the CIA tries to get congressional action and he can't, what can a lone scientist at Ohio State University do?

VALLÉE: What happened in the early days of NICAP? NICAP now is not very active, but in the early days they were. Didn't they put the Air Force under a lot of pressure?

HYNEK: Oh, well, my impression of NICAP was completely soured and prejudiced, perhaps because I never had any direct contact with them; all I knew about NICAP was through Blue Book, and they were painted to me as a bunch of crackpots. Keyhoe was presented as a scoundrel and a mountebank, and all I remember was that, time after time, when Keyhoe seemed to be getting a little ascendency and calling for congressional investigations, suddenly there was a big flurry at Blue Book, key congressmen were called and told not to pay any attention to him, so there was a real counter-offensive mounted ... and after ... I wasn't sharp enough to see that this was a highly political move. I really thought then that NICAP was a bunch of nuts.

VALLÉE: And the NICAP people never bothered to contact you to find out the extent of the problem from your standpoint?

HYNEK: No, NICAP certainly didn't. Instead of trying to fight city hall all the time, if they had gone ahead quite independently and simply forgotten the Air

Force, they could have done something big. They could have said, forget the defense people, we'll surround ourselves with some good scientists here, and we'll investigate the best cases. They were getting money, you see, they had enough money to do this.

VALLÉE: They had more money than Blue Book did.

HYNEK: Actually, for a time, yes. Blue Book boasted that they had access to all the scientific facilities and all the technical facilities of the entire Air Force, which on paper was true but . . .

HASTINGS: But the commanding officer took the train. . . .

HYNEK: Yes, but there was one characteristic of Blue Book, and Hardin was a good example, he latched onto this; when there was an obvious case of reflection of the sun, a reflection on ice crystals that were below an airplane for example, and this reflection traveled along, any natural explanation they could get like that, they played it up and *they used that in their briefings*. That one picture was used at least a dozen times in briefings in Congress and so forth, "See this is what was reported as a UFO!" but *they never presented one of the unsolved cases*.

HASTINGS: It was the only thing they had to prove they were doing their job. . . .

HYNEK: Yes, they couldn't bring in anything unidentified because they weren't doing their job if they hadn't solved it. It was a game. It was how many goals could you score against the other guy. You closed up the cases you solved, made a beautiful box score out of that and just said the others are psychological.

The Blue Book Fallacy

HASTINGS: I see a rationalization there that says, *we* went through X amount of effort and we solved the case, and we found it was a reflection of the sun . . . well, that leads you to the feeling that if *you* haven't solved this, what you need is X plus some effort and you will have it solved, too.

HYNEK: This was exactly the point, and the fallacy of Blue Book was really apparent. Their statistics, for instance, were really false. They boasted that they had only 2 or 3 percent unknown and based on that they would argue in a very interesting circular-reasoning way: We have solved everything but 2 or 3 percent (hence it follows that if we really tried harder, if we had a little more data in these cases, we'd have solved the other 2 or 3 percent; after all, it's only 3 percent!). Then they would complete the circle of reasoning: If that is the case, no point trying! But it wasn't 2 or 3 percent. It was 20 percent, and I was a little dumb in those days. I don't know why; I guess, sometimes a person's scientific training can work against him rather than with him, because I always was looking for the absolutely *hard core* data. I wasn't going to say a thing until I had the perfect case; bang, bang, it would be the landing on the White House lawn. I was always waiting for that, and I wasn't putting together the bits and pieces of the unidentified cases.

VALLÉE: I remember the first time we met in Chicago late in 1963, when I showed you some statistics based on the French data. I asked you if you had read Michel's book. You said yes but there were no data like that in the United States. It's true that the Air Force wasn't getting that kind of data. It was there, it was there in NICAP's files, but NICAP was too afraid of it to publish it, and there's a lot in those files that hasn't come out yet that is of that nature. If it had come out then ... this is a subject of great puzzlement to me ... why this sort of thing wasn't known in the United States. On the one side, you had the Air Force, with the picture you described; on the other side you had Adamski and the cultists. *And there was nothing in between.*

Now in France in those days, we didn't have the Air Force and we didn't have Adamski, but we had people who were seriously gathering stories without making assumptions about them, just gathering them and publishing them ... as interesting stories. And I remember our discussion there when you said, "Well those cases that Michel quotes, I read them like you read books about ghosts ... not having the background, I have to say, maybe this happened and maybe it didn't."

HYNEK: This is true, we didn't have those things. We had some landing cases, but time and again, what today I would regard as a good physical trace case they would have marked "hoax"; it couldn't be real, it just had to be a hoax. If somebody said they had three triangular markings, then the kids must have put those in there to make a good story out of it; that was just standard technique! And if it weren't a landing, this was simply automatically labeled—I saw it with my own eyes—"psychological." You just have to realize that when you talk about people with blinders, persons who have only learned how to do something one way, they are simply not open to any other suggestions. And the lieutenants, the sergeants I had to deal with there ... it's hard to convey the box I was in! Never once did I have a truly scientific conversation with any of the people. You couldn't look open-mindedly at a case and say it might be this or that; no, the guy had to be a nut, "It can't be that sort of thing ..." or "They said it was like this, but you know how wrong human perceptions can be."

If there was any chance of getting out of it, they got out of it! Coming back to those statistics, the reason that those 2 or 3 percent were dead wrong is that, first of all, the cases they honestly felt did not have enough data or information were marked insufficient data, and they regarded those as *solved.* Then, of course, even more heinous than that was the fact that if this UFO had a light that moved—well, airplanes had lights that moved, and so in their minds it must have been an airplane—and they dropped it into a category such as "possible aircraft." At the end of the year, when they made up the statistics, the "possibles" and "probables" were dropped, so even a "possible balloon" became a "balloon." That's how they got the residue down to only 2 or 3 percent. All that time, however, they never evaluated any case as "possible UFO"!

VALLÉE: They went from "possible balloon" to just "balloon."

HASTINGS: You could do that on the computer. [Laughter.]

VALLÉE: Yes, the computer is very good for it!

HYNEK: Blue Book was looking to make brownie points for the Pentagon. I was always rather naive about how military matters worked. The guy who was in charge at that time, the big cheese in charge of the whole area there, had been a real big shot in Japan. He had all sorts of servants; I never met him, I was too far down the line, but I'd see his signature. His move to Wright Field he regarded as the greatest sin perpetrated since Judas. It was terrible that he should be treated that way. He didn't care about his new job and particularly this nonsense about UFOs. He simply rode roughshod over it. He didn't want UFOs, he didn't want Blue Book, it was a damn nuisance to him! He was dissatisfied. He was completely disgruntled with his own job and so the people who were in charge of Blue Book just really had a hard row to hoe. Even if a man in the project felt that there was something genuine to it, he couldn't have said it. He would have been ridden out on a rail by the top brass at Wright Field. So the people who want promotions and retirement, they aren't going to rock the boat; why should they? They would have been completely out. So it was a sad situation all the way through.

One thing that Jim McDonald could never recognize was the sort of box that I was in. And, of course, the first reaction that comes to a person is: as a self-respecting scientist, why would you ever stay with something like that? I definitely was intrigued. I wanted to be there. I enjoyed coming there each month and sitting down, reading the reports that had come in and keeping my own counsel.

Today, I wouldn't know what the Air Force files are like if I hadn't been there. I sort of swallowed my pride and recognized everything was a game and there was no chance of having a scientific dialogue with these people. I wish there could have been scientific dialogues.

There Couldn't Be Another Condon Report

VALLÉE: In a way, the changes have seemed very slow, but you can look at it from another perspective and see that, in fact, opinion has changed very fast. Just a couple of weeks ago, you and I were talking about UFOs to a group of scientific colleagues, and we could present that kind of data, including landings, we could even discuss occupants and be taken seriously. Doesn't that mean that the change has been occurring fast really in the past few years, that there is something happening? We now accept certain concepts that enable us to think about things that were impossible for Ruppelt or for anybody in 1950 or 1955. At the time, one couldn't recognize what cases were important to follow up, even though you had the intuition, you didn't have the logic that would enable you to define that kind of criteria. Now those criteria are available and more scientists are ready to recognize them and study them. To that extent your attitude, and your saving the data, has contributed to a very rapid process of creating new concepts. Now doesn't that have some consequences in terms of scientific knowledge? Nobody could stop this completely again. There couldn't be another Blue Book. There couldn't be another Condon. The climate has changed dramatically. One need only point to the recent scientific meetings on UFOs— conferences that would have been literally unthinkable a decade ago.

HYNEK: There couldn't be another Blue Book. If there were another Blue Book today, with the data I've got and the things I could show, I would not be afraid. Having a tenured position and being close to the end of the line, I wouldn't be at all afraid of pulling that out and testifying. For many years, the Blue Book operation had very low priority as far as the Air Force was concerned. And it was sloppy, just kid's stuff, actually. If there had been some way of bringing in the NICAP or APRO data, if they had had honest-to-God scientists, with funds to publish a really good "UFO Evidence," the picture might have changed, but scientists were unaware that the amateurs were getting the good data they were. When Keyhoe wrote about it, it seemed terribly sensational.

HASTINGS: Where did the amateurs get the funds?

VALLÉE: Membership fee of five dollars initially (now it's ten dollars). At one point, they claimed they had ten thousand members, that would have come to fifty thousand dollars a year for NICAP alone.

The Swamp Gas Incident

HASTINGS: What was the progress of your changing attitudes? I mean, you mentioned at one time you were vacillating. First you were skeptical, you know; what were those changes you went through? And how did they come about?

HYNEK: I think I finally decided that things were serious much too long after I started. I had internal misgivings about Blue Book policies as early as 1953, and I expressed them to my colleagues when I had a chance.[4] It wasn't until after the

---

[4] See Hynek, *Journal of the Optical Society of America*, July 1953.

"swamp gas" incident that I said, "I've had it! This is the last time I'm going to try to pull a chestnut out of the fire for the Air Force." In the case of the "swamp gas" incident, it could have been totally different, Jacques, if I had simply, out of my own pocket, paid your and Bill Powers's way, or if you had somehow come along. I think you would have advised me differently at that point. I was there alone in a rather circus atmosphere. I should have said, "Look, we cannot conduct a scientific investigation in twenty-four hours. I want some colleagues in here." Today, I would do it quite differently. I'd say, "I want to bring Jacques Vallée in, I want to bring in Bill Powers, I want to bring in Fred Beckman. I want to get them over here and set up a team to work on this thing for the next several days." But I didn't.

VALLÉE: The thing happened fast. I remember that you said, "Hold on and I'll call you if I want you here." Bill and I were ready to fly over there. We had a private airplane lined up to fly to Michigan, but it was all over in a few hours.

HYNEK: This was the case that really triggered public passion and brought on the Condon Committee in the long run. A farmer named Frank Mannor and his son had reported a landing in Dexter near Ann Arbor in March 1966. I had just come back from New Mexico with a broken jaw—my jaw was wired, and I was not feeling well. I think if I'd been in top-notch health, I would have been thinking straighter. In any event, there was a mess of reports of landings and brilliant lights, also a hell of a lot of faint lights.

I think, another thing, if I'd been able to stick with the Frank Mannor case and there hadn't been anything else, then it would have been different. But then the Hillsdale girls came into it. They were students at Hillsdale College, which is about forty or fifty miles or so from Ann Arbor and Dexter. These girls had reported lots of lights flickering around the arboretum. The lights were faint, they were confined to a swampy, moist area. They were not bright. The thing that really got to me was when the girls said they had to put the lights out in their rooms to see the lights outside.

VALLÉE: Neither the lights nor the girls were bright! [Laughter.]

HYNEK: At any rate, when I had to have a press conference, I really did the wrong thing—it was Huntley-Brinkley and all that.

VALLÉE: If it had taken place one year earlier, the report you gave to the media would have been accepted as "the scientists have explained the sightings." But the climate had changed drastically in that one year. You underestimated the impact of the work we had all done together between '63 and '65.

HYNEK: Well, we were getting tired of the Air Force's pat explanations, that's for sure. Oddly enough, I really think *those small lights may very well have been swamp* *gas!* I went into the history of swamp gas and the chemical reactions involved. Foxfire, will-o'-the-wisp, does occur and has been described pretty much the same way. I was completely honest in the press conference. I pointed out that I was talking only about the faint lights that had been seen. Those could have been explained as swamp gas. I deliberately excluded what I thought at that time to be fanciful tales of bright lights chasing policemen in the area.

VALLÉE: The case in Milan?

HYNEK: Yes, in Milan. That photograph is another thing that set me off; I took one look at that photograph all the papers had published, and it was obviously the moon and Venus rising. Now, what happened is that even though Congressmen Gerald Ford and Vivian were on opposite sides of the fence, politically, they both decided that they'd had enough. They resented the ridicule the State of Michigan was getting, being called the "Swamp Gas State." The cartoons, the editorials, the stuff that was written at that time about it . . . I was lampooned and had all sorts of cartoons. One showed little men holding up guns and me roped to a flying saucer, and they were aiming at me saying, "Swamp gas, huh?"

HASTINGS: So people really took it as quite an inadequate explanation.

HYNEK: Yes, it was a beautiful reaction against the Air Force and me personally. They'd had enough, and here to them was an obvious attempt to whitewash something again and, to that extent, it served a purpose. This led to the congressional investigation, which through many devious ways, finally led to the Condon Committee.

VALLÉE: Isn't what we have here the kind of thing that Arthur was saying recently about *belief structure changes*? A new belief structure has to be constructed, and from hindsight now you can say, "Well, if I had known in 1955, I would have done things differently too, but there was no way." I think I can testify to that by the atmosphere at the observatory, for example, when I first met you in '63. And then, certainly, by the atmosphere at Wright Field. I remember the meetings with Colonel Quintanilla and Sergeant Moody, when they came over to my house in '64 and I showed them some data from Europe, and I said, "If you look in your files, you can find reports just like that." They just wouldn't believe it. They didn't even know their own files. There was a climate of just total disbelief.

Looking for the Haystack

VALLÉE: I find myself now looking at data in a very different way from how I was looking at it five years ago. I've had to reconsider many of the things I said earlier about the contactees. Some of the contactees may be telling the truth. Ten years ago, I said, "Well, those people obviously are lying." I wouldn't even have stopped ten minutes to listen to them. So I think what you say is really illustrative of a really big, large-scale revision of the intellectual concepts.

HYNEK: In the fall of '66 was the real time I changed. I said, "I'm going back and relook at this material with a different viewpoint, no longer assuming that the chances are strong that it's all a lot of junk. I must take the viewpoint that after all these years, the data may be genuine, as poor as it is (because it is not hardcore data, not the sort of data that a physicist wants). I'm just not going to continue to call all these people liars, deluded brains, and so forth, even though some of them can't tell Venus from a hole in the ground; they're people whose testimonies would be accepted in a court of law in any other context. By what right do I have to continue to doubt their words?"

VALLÉE: Isn't there a danger of a backlash the other way? Once the dam of rationalism, of skepticism breaks, suddenly people are willing to believe *anything*. People are going to go out and look at Venus and the moon and say, "Look at the Flying Saucer!" They won't even believe a scientist who *proves* to them that it's Venus or the moon. It goes completely the other way. I see that tendency in the works of a freelance writer, a friend who has written several good articles and books on the subject; he has described the passion of some of the local waves where people are literally going berserk, seeing lights in the sky that are just ordinary, explainable lights in the sky, but once the pattern has been broken, they panic and think, "Maybe this is a UFO, and it's coming to get us!" Once that rational debate is suspended, then people will accept anything. Isn't there a danger of that?

HYNEK: There's a real danger of that, sure.

VALLÉE: Isn't that even greater than the other danger, which is not even to consider it?

HYNEK: Yes, I saw it in my own life. I was tempted a number of times in the '50s to suspend disbelief and jump over, but as I said before, with scientific training, you have to prove everything and have everything in charts and graphs; also, I was constantly coming up against sheer stupidity. Some of the reports we got at Wright Field were just, you know, "My God! How dumb can people be?" that attitude. When we saw that in case after case, it just sort of carried over and smeared over, whitewashed or blackwashed, the really good cases. You just couldn't take the good cases seriously when you saw all this crud, all this noise, all the junk. And then Adamski and the cultists!

At times, I said to myself, "Oh, to hell with all of this business! If there's something really here, it's so buried that there's not a chance to find it. It's like looking for a pearl in a cesspool—you run through a hell of a lot of stink and muck to get at something, and you're not sure the pearl is there."

VALLÉE: Ira Einhorn, a poet in Philadelphia, has a nice expression for this. He says, "It's like the needle in the haystack; we have the needle and we are looking for the haystack!" [Laughter.]

HYNEK: I'm not quite sure I understand that.... Well, I also remember the first time I met Jim McDonald. He thought I was so completely dead wrong and so obtuse! He came to visit me at the office and he pounded on the table and he said, "Allen, how could you have sat on this data for eighteen years and not let us know about it?" And I said, "Well, damn it, what data? Large parts of the data are sheer nonsense!" But at the same time I had a tremendous feeling of relief. Here was another scientist, the first one I had ever come across who was established in his field who actually was taking the UFO problem seriously. You have no idea what a relief it was, like taking off a pair of very tight shoes, to have a sympathetic listener. There was not one of my colleagues interested in it, not even on my own staff. All those people were absolutely convinced I was a nut. So, the very fact that Jim McDonald existed was a relief, even though he thought I was wrong and should have gone and pounded on the doors of the generals. I said, "But Jim, you weren't there. You don't understand. When you have these senior scientists like Robertson, a Nobel Prize winner, Luis Alvarez, another Nobel Prize winner, all saying it's nonsense, you don't get an assistant or associate professor going up and pounding on generals' doors and saying, 'Look, listen to me'—you're a voice crying in the wilderness."

I still think the smart thing to do was to just bide my time—wait around and sooner or later the climate would change, the same way as Wegner did on continental drift, and other people whose names I don't know on acupuncture. Think of all the things in science that were vilified and ridiculed! Well, Bernard Shaw said that all great ideas start as heresies, and this flying saucer thing was really a heresy. Also, the term *flying saucer* was most unfortunate. If it had been introduced as some other term, it would have been less open to ridicule. If they'd called it an anomalous observation or strange optical phenomenon, or something like that—but *flying saucer* immediately gave the public a handle and an excuse for ridicule.

HASTINGS: And the fact that it had so much availability to the public and so little to the scientist as a phenomenon.

The European Scene

VALLÉE: To what extent was your mind changed or your belief changed by discussing the European cases?

HYNEK: Well, I must confess, when I read Michel's book,[5] I said to myself, it just can't be—these stories of a man coming home, putting his car away in the garage, and looking up and seeing these cigar-shaped objects. Well, it was a foreign country, it wasn't happening here in the United States. It was something strange happening in France. I didn't think the French were crazy or anything of that sort, but all I knew was that we weren't getting that caliber, or that kind of data in the Blue Book files. We found occasional cases like that, but we didn't have a pattern.

The thing that really impressed me when I visited Michel in his apartment in Paris was that he had newspapers from every little town in France, literally from the floor to the ceiling in this little room. He had had to patiently go through and cull them out. The reports didn't appear very frequently in the Paris papers. Occasionally they had one in *Figaro*, but generally it was buried in small provincial papers, and he painstakingly culled that out. The wave of UFO sightings was truly an impressive one—many hundreds of reports over a period of two months, a sizable percentage of which came from credible witnesses. Yet, the whole thing would have been lost to history if it hadn't been for the painstaking and pioneering work of Aimé Michel. Well, that really impressed me. And I saw that, well, okay, there really was this wave in France, and just reading Michel's book the first time, I didn't get it, I wasn't convinced.

VALLÉE: We spent two days in 1963 discussing statistics based on that, and the correlations between patterns in France and other countries. To what extent did that change the assumptions?

HYNEK: The French cases were different from the cases in the Blue Book. We had so much crud in the Blue Book! It wasn't until '66 that I decided I had to revise my views and take a new look, literally a new look at the whole thing from a different vantage point. Then, suddenly, things began to make sense to me. But look at the time it took, from 1948 to 1966—eighteen years. You would think that I would have certainly seen the value of the material, but being alone, in this country at least, I thought I would have to have overpowering evidence before I really let the presses roll.

It's a little bit like an experience I had in a washroom at Ohio State. A group of psychologists were having a meeting at the University, and it so happened that Rhine[6] was in the cubicle in the men's washroom next to me. One of the psychologists was at the washstand washing his hands, and the conversation went something like this—the guy at the washstand was saying to Rhine, "Well," he

---

[5] Aimé Michel, *Flying Saucers and the Straight-Line Mystery* (Criterion, 1958).

[6] Dr. J. B. Rhine, author of several books on ESP.

said, "your statistical results are sort of impressive. In my own field, if I had one-tenth of the data you presented, I'd believe it, but I'd have to have a hundred times more from the ESP field." That summed up the feeling I'd had. If it had been in astronomy, if I'd had one-tenth the data to prove that a certain star had the wrong parallax or something of the sort, I could have gone and written a paper on it. But here in a completely hostile atmosphere, I thought that I had to have a thousand times as much data before I could go out and say something.

VALLÉE: When our "Invisible College" started meeting in Chicago at the end of '63 or early '64, that provided the kind of framework where suddenly it was okay to speculate about the reality of UFOs. I remember that we used to call witnesses, for example, when we were together on Friday night and Saturday night. There were mathematicians and psychologists in the group, and we would call witnesses sometimes for forty-five minutes to follow up on some of the Air Force cases. I think that had something to do with changing the minds of all of us.

HYNEK: I'm sure it did.

VALLÉE: Suddenly, it was possible to talk to someone else about it; it was an environment where you could say anything. You could say, "Hey, wouldn't it be funny if those turned out to be really from Mars!" You could joke about it and laugh about it.

I remember when I first met Allen. I knew Aimé Michel very well. I had the other side of the picture, because I knew the French and European data. But I assumed that if there was anything to what we were thinking in France, then the U.S. Air Force must know about it! The Air Force must have all that evidence and more. So, meeting Allen and having this kind of discussion was really a revelation to me. We already had statistics that showed that there were common patterns. I had compiled statistics on the time of day, and when the landings occurred. Statistics on the large objects releasing the smaller objects, the patterns of contact. When we started putting the two frameworks together, it started to click. To me, that was a turning point.

The Frustration

HYNEK: There was one thing we lacked. The thing we haven't really brought out here was the frustration from lack of follow-up. The Air Force had a process for finding a possible solution and that was it.

For instance, there was this case in which a young couple said they'd been followed by a UFO and they came home nearly hysterical. The Air Force took a look at it, I think it was Sergeant Moody who looked at it. Well, the moon was full that night, near full, and it was in the east. They said the thing had come from the east, therefore the Air Force said they must have been chased by the moon! They must just have looked in the back and had seen the moon in the rearview mirror—this was their solution. No attempt was made to go and talk to these people, none at all. They'd arrived at the solution that it was the moon.

Oh, I don't know how many times I asked to have a case reopened or looked into, or followed up. When they came in from some air base, say in Texas, I would say, "Well, let's call them back and let's find out. Let's get this straightened out," and so forth. Have they talked to the witnesses again? Were there any other witnesses? Never! They would simply not call them back. They didn't want to bother. For one thing, Blue Book did not have any clout. Being generally headed by a captain or so, the captain couldn't command a major or colonel at some air base to do something. It had to go as a request.

Now, if they'd had a colonel or a general in charge of Blue Book, then he could have picked up the phone and called anybody at any of those air bases and they, by golly, would have done it! But Captain Hardin, for instance, even if he'd been so inclined, couldn't have gotten any cooperation. A mere captain telling a colonel at such-and-such an air base to do something? You don't do that sort of thing in the military! Very sad, and sometimes I felt that I had gone as far as I could and I thought, well, if they're going to be so damn dumb and not follow these suggestions, it's just too bad for them. I made suggestions about cross-indexing, suggestions about follow-up. No. It was just an absolute impasse.

It's hard to describe my feeling of frustration about Blue Book. And yet, I enjoyed going there and reading some interesting cases. It was just, I got that feeling that, well, at least privately, I'm getting something out of this. I'm enjoying this inwardly. As if reading Agatha Christie were forbidden and you would grab it and then hide in some room and read it! That's the feeling that I had. I would come across an interesting case, and do you realize I could not discuss it with people? They simply wouldn't discuss it. And you have, I forget the lieutenant's name we had ...

VALLÉE: Lieutenant Marley?

HYNEK: Yes, you couldn't sit down and tell him, "Well, this is interesting. What do you think might have happened here? Do you think there's a possibility of this?" or "I don't think they're lying, I think they're good, solid, people. Why should they say something happened?" You couldn't have that sort of discussion. They had spoken. They had decided it was a balloon! I think, also, one of the rather amusing things was the reaction at Blue Book when a person reported a flying object. Just to show you how ludicrous the whole thing was, they said, for example, that they saw a craft move from a northwest to southeast direction. Quintanilla would look blankly at me and say, *"Well, we checked the radar and they had no aircraft there. Therefore, those people couldn't have seen it there!"* They had that attitude. There were no aircraft there. These people reported an aircraft. They never asked themselves whether these people had reported a UFO! The UFOs didn't exist! If people saw something, it just had to be an aircraft. Well, their radar said there was no aircraft, therefore, these people are crazy.

VALLÉE: Has it occurred to you that the fact that *you*, of all people, were in that position of seeing this, of witnessing it, of saving these data, was more than a coincidence?

HYNEK: In what context?

VALLÉE: You had the kind of mind that was attracted to it. You had an astronomical background; you had the knowledge necessary to separate the signal from the noise and so forth, but you also had the curiosity of someone who was almost mystically inclined. You were able to recognize something that went beyond the limits of current science.

HYNEK: Yes, but it wasn't due to any selection, unless you describe it as sort of a paranormal coincidence.

VALLÉE: That's what I mean.

HYNEK: Yes, well . . .

VALLÉE: Almost in a psychic sense.

HYNEK: Almost a psychic sense . . . because they could have had any one of a hundred consultants. I often say that I was the handiest astronomer. There were several other astronomers at Ohio State that they could have picked, but they recognized that I was sort of a tame astronomer. I was, at least apparently, going along with the game. I wasn't fighting. Had I been fighting, I would have been kicked out. As long as I wasn't fighting them and calling them damn fools for not looking at the data better, I was tolerated. I was all right, and they frequently used my name. "This has been examined by Dr. Hynek, and he is an astronomer," and so forth. In one sense, I was definitely prostituting myself.

HASTINGS: It's curious though, at the same time, that if it hadn't been for that, you wouldn't have had the data, we wouldn't have the data. It simply wouldn't be here.

HYNEK: We wouldn't be here around this table talking right now.

HASTINGS: The only thing is, if you had played a better game of it, maybe you would have had more. Do you feel anything like that?

HYNEK: Well, I could have been playing more of a game, but. . . .

HASTINGS: But you didn't know you were playing a game at the time, so you really can't hold that against yourself.

HYNEK: I could have been sharper, and I could certainly have kept a hell of a lot better notes. There were many papers that went through my hands the significance of which I didn't recognize. If I had been more of a cloak and dagger man, I could have done more. I have to have something right in front of me like that, then I say, "Oh, my God, yes!" But I don't look under the table or under the bed to find it. And I'm sort of learning to finally, but, on the other hand, I saw the picture that some UFO enthusiasts presented which in many respects was galloping paranoia, and I certainly didn't want to be that way.

The Dangers of Cultism

HASTINGS: It's very good, you see, to have people like Adamski and the like because they hold down that end of the spectrum and gather those kind of people around them in one place.

HYNEK: But think of the damage they did.

HASTINGS: Well, we haven't found ways of handling those attitudes socially.

HYNEK: You see, the very fact that Adamski and his cult existed and people said they were taking trips to Venus, and Adamski was saying that there were people on the other side of the moon and they had this Big Rock Convention, and people selling hair from the Saturnian dogs . . .

VALLÉE: You immediately became associated with them if you only used the expression *flying saucer*. In the mind of the more educated public, you'd be in the same box with the worst crackpots, because people would identify you with them.

HYNEK: Sometimes it seemed to me that we could almost adopt the hypothesis that intelligences were coming here secretly to take a good look at the earth. They could have deliberately, somehow, picked and energized people like Adamski, to completely muddy the waters! It would be a beautiful counter-espionage plot.

VALLÉE: Even now, whenever there is an article in some popular magazine on UFOs, the Adamski picture is printed in it. That is the one picture that is associated all the time with the subject. It's not just that those magazines want to ridicule people who talk about UFOs. It goes beyond that.

Just a couple of months ago, a well-meaning fellow in Berkeley put a panel together to discuss UFOs. He wanted Jim Harder there, and I was invited too, but I saw that the leaflet announcing it had Jim's name and my name and there also was the Adamski picture! And I said, "Why did you do that?" I told him that I was cancelling, that I wouldn't be there. And he said, "Why not?" Well, I was sorry, but the image was all wrong. I couldn't get the message I wanted to get across, I wanted to talk about psychic aspects of UFOs. I couldn't do it with that image. And he said, "But I didn't mean to do anything bad. I used that for *graphic impact!*" Even now, people associate this immediately with UFOs. It's impossible to differentiate. If you talk about flying saucers, you're identified with the most extreme of the cultists. It's very hard to draw a line in the mind of the public.

HYNEK: Well, look how that happened to me at the press conference in Detroit during the "swamp gas" case. All those reporters and photographers around, suddenly somebody just took that Adamski picture and shoved it into my hand like this, and there I was, the photographer's flash went off, and there I was. This was the last picture in the world I would have wanted to be photographed with, and some reporter zoomed it into my hand, "Well, what do you think of this?" It's a natural reaction that when somebody hands something to you, you take it! And then that same picture appeared again and again. It seems like some sort of a plot to ridicule the whole thing.

Behind the Scenes

VALLÉE: Adamski had two different kinds of language. On the one hand, he was a very sane man, and on the other, he was saying some things that were absolutely preposterous—as if he *wanted* to be ridiculed.

HASTINGS: That may just have been his psychological makeup.

HYNEK: Like Condon. Condon was a perfectly sensible man and a very astute scientist, and look what he would go around saying—completely violating all the principles of science by going out and talking one year before the report of his committee came out, saying in effect, "It's all a lot of bunk, but, of course, I'm not supposed to say this for a year."[7] Now, what sort of responsible statement is that from a scientist?

---

[7] The following news item appeared in the Elmira, NY, *Star-Gazette,* January 26, 1967, very shortly after the Condon Committee first convened.
>   MOST OF UFO SIGHTINGS EXPLAINABLE—SCIENTIST ADVISES CORNING AUDIENCE. Corning—Unidentified flying objects "are not the business of the Air Force," the man directing a government-sponsored study of the phenomena, Dr. Edward U. Condon, said here Wednesday night.
>
>   In an hour-long rundown on the government's interest in the field and the recollection of some baffling and spectacular claims by UFO "observers," Dr. Condon left no doubt as to his personal sentiments on the matter:
>
>   "It is my inclination right now to recommend that the government get out of this business. My attitude right now is that there's nothing to it."
>
>   With a smile he added, "but I'm not supposed to reach a conclusion for another year."

VALLÉE: You could make a cloak and dagger story about Condon too, if you wanted to.

HYNEK: Well, of course, one or two members of the Condon Committee have. They think the whole thing was rigged.

VALLÉE: You and I certainly never saw any evidence that it was rigged. When we briefed his committee in 1967, the whole atmosphere there seemed to be very open-minded.

HASTINGS: It's pretty clear that all the people on the staff were not in agreement, and they were simply saying what each of them wanted.

HYNEK: I think many of the people on the Condon Committee were trying to do an honest job. They were sincere. I know this from my conversation with them, both during the life of the project, and later. It is also brought out by Dave Saunders's book,[8] which, incidentally, is today a collectors' item. It should be reprinted. And in Sturrock's scholarly critique of the Condon Report[9] he reflects the idea that at least some of the men on the Condon Committee were quite sincere, but generally did not have Condon's ear. Likewise, in the report of the AIAA (American Institute of Aeronautics and Astronautics) Subcommittee on UFOs, the same idea is expressed—Condon was obviously not aware of what his own staff was turning up. At least, that is the most charitable explanation.[10]

VALLÉE: I was very leery when the request came from Condon to turn over to them my computer files. You know that the catalogue that preceded UFOCAT started with three thousand punched cards that I sent to Dave Saunders, and I was very leery of doing that because there was the suggestion all the time that it would be used negatively by Condon. I saw a piece of paper with the heading: *computer statistics,* five thousand cases; three thousand from Vallée and two thousand from others! Okay, I suspected that maybe they were going to take all the statistics that were presented in *Challenge to Science* and issue a negative report by biasing the data.

---

[8] *UFOs? Yes!* Saunders and Harkins, Signet paperback Q 3754.

[9] Peter Sturrock, "Analysis of the Condon Report on the Colorado UFO Project," *Journal of Scientific Exploration* 1, no. 1 (1987): 75–100.

[10] J. P. Kuettner et al. "UFO: An Appraisal of the Problem," *Aeronautics and Astronautics* 8, no. 11, (1970): 49–51. In several parts, it states specifically, "not all the conclusions contained in the Report itself are fully reflected in Condon's summary," and also, "Condon's chapter, 'Summary of the Study,' contains more than its title indicates; it discloses many of his personal conclusions."

They could rig the statistics to come out any way they wanted by adding an appropriate number of hoaxes and errors. They could say, "Those statistics include, and therefore supersede, all the data from Vallée and others because we've got his data as a subset of ours." So, an atmosphere was created that was very unhealthy. I did turn over my data to them because I had great trust in Saunders, personally, but I felt very disgusted with it and, frankly, I left the country as a result of that and went back to Europe for a year. I just couldn't stand it any more. That exposure to big science . . . I thought, "Hell, if this is science, then I don't want to be a scientist!" It had a very, very deep effect on me at that time, in 1967.

HYNEK: Condon, to some extent, could almost be excused; he was the figurehead. He was busy with other things. He knew he would only spend part of his time—he gave the running of the thing to others.

One thing really astounded me. Were you in the room with me at the time when all the chapter headings of the Final Report and all the conclusions were written on the board? In the first week of the Condon investigation, they were writing the Final Report! They had started from the basis that the results were going to be negative, there was no question about it, and they were already putting down the chapter headings as to how they would present the thing, how they would present their negative results. He was doing that. And, of course, in the famous memo by Robert Low that was certainly brought out.[11] But I didn't know about that memo until much later. Well, maybe all's well that ends well. We wouldn't be having this fun here if the Condon Committee had come up with a positive report!

VALLÉE: When you asked Condon what he would do if he found positive evidence of the existence of UFOs, what did he say?

HYNEK: The response from Condon was that he would take it to the president. And it would undoubtedly be classified, which I suppose is sensible. I mean, if you suddenly found out that it was absolutely genuine, that we were faced with

---

[11] It was on August 9, 1966, that he wrote the memo that later led to so much trouble. Addressed jointly to Deans Thurston E. Manning and Archer, the memo juggles the pros and cons of the University of Colorado's possible acceptance of the Air Force proposal. In it, Low reviewed the reservations held by various persons with whom he had consulted, but nevertheless concluded that it could be to the university's advantage to take on the project. Then he wrote: "The trick would be, I think, to describe the project so that, to the public, it would appear a totally objective study but, to the scientific community, would present the image of a group of non-believers trying their best to be objective but having an almost zero expectation of finding a saucer." From Saunders, *UFOs? Yes!* p. 129.

something like that, it would be similar to taking the radar indications of Pearl Harbor seriously. If we found positive evidence, then the security of the country would be in jeopardy, perhaps, and I think Condon would have been right.

HASTINGS: Somebody told me that a lady was in to see Humphrey one day when he was vice president, but he wasn't in his office, so she looked on his desk and saw a report entitled something about UFOs from the U.S. Air Force. I can't imagine her opening it and looking, but she did, and its conclusion was that they existed and they were apparently extraterrestrial. And when Humphrey came in, she said, "Mr. Vice President, this is incredible, marvelous, most interesting, and very exciting and when is it going to be announced?" And Humphrey said, "Never! It would be too terrifying to the American people, and it should never be released."

VALLÉE: What she saw may have been some official-looking report from NICAP. Is she certain it came from the Air Force?

HASTINGS: I don't even remember who told me the story.

HYNEK: With this kind of thing, we have to clearly say: these are rumors. We haven't the slightest idea whether they're right, but for what they're worth, here they are!

The Credibility Gap

VALLÉE: Isn't science going to have a credibility gap in the matter of UFOs?

HYNEK: What do you mean?

VALLÉE: I just think that, on one hand, we have the recent Gallup Poll observation that 51 percent of the American public believes in UFOs. Yet, on the other hand, the proportion of scientists who would openly agree that UFOs are real is still very small. This leads me to believe that there is a credibility gap in the sense that the public believes that something exists that science doesn't want to study.

HYNEK: It isn't that the scientists don't believe or are against it so much ... but it's like discussing sex openly in a Victorian drawing room ... the Victorians knew that sex existed and they went to the other side of the tracks and all that, but in polite society it wasn't mentioned. Well you have that same situation here. The UFO is not supposed to be mentioned in polite society and yet in the public mind it is very much there, and that does create a tremendous credibility gap.

VALLÉE: My feeling is that if things continue to develop the way they have been developing in the last year or so, scientists may find themselves caught between the public and the government. The government wanting a study of UFOs, the military wanting a study of the UFOs, the public wanting a study of the UFOs, and the scientists suddenly cut off from both the popular base of science and their sources of funding. This is very striking in France where the rationalist scientists are telling the government there is nothing to the UFOs, and the Minister of Defense is saying that "the facts show the phenomenon to be real." Not facts from the scientists, mind you, but from military pilots, from military radar, and from the police. I think the same thing might happen in this country.

HYNEK: Well, certainly acupuncture didn't get any credence because of the AMA. It was from the patients themselves, people who said it worked.

VALLÉE: I can think of two other instances where there were similar gaps. Consider these two major developments in this century, the artificial satellite and the computer. In both of those cases, many scientists were against developing the device. They thought building it was impossible when it was first proposed, and they tried to stop it when it was started. A number of influential astronomers have said we shouldn't be doing space science, we shouldn't be launching rockets, we shouldn't be launching satellites. Every time astronomers were asked what benefits could come from satellites, they said, "None! What we need is bigger telescopes on earth, and more telescopes on balloons; we need more balloon astronomy!" But the usefulness of an observatory on the moon or in space, they couldn't see it. Do you remember the Astronomer Royal of Great Britain saying in 1957 that "space travel is utter bilge?" Yet astrophysics has been revolutionized in the last twenty years because of satellites.

The same thing with computers. Computers were developed because the banking industry wanted them and the Pentagon wanted them, but most scientists (outside a very small nucleus of people in numerical analysis who had to solve fancy differential equations) had absolutely no interest in them. I can remember, after I got my master's degree in astrophysics, telling my adviser I was interested in computers and his saying, "Don't waste your time with courses in programming because that is a fad that is going to disappear. Those things have no long-term scientific application!"

HASTINGS: It seems to me that your Center could try to bridge this credibility gap. It might very well have a magazine or journal for professionals in the field that would be read by your colleagues in the scientific area. It's one way of showing you have a product that you're really working with, and it might be very useful because in any of these interdisciplinary kinds of topics, your communication is not the same as in a professional field, where you have a journal and conferences. You have to do it almost sub-rosa.

VALLÉE: I've always been struck by the fact that in the *Encyclopaedia Britannica*, the term "unidentified flying object" falls between Unicorn and Unified Field Theory. In itself, it's a good definition for a UFO: "Something which is between a unicorn and a unified theory!" [Laughter]

HYNEK: Both are mythological objects!

8
The Literature

Personal Motivations

HASTINGS: Let's say I'm a person who knows nothing other than what I've read in a few newspaper accounts about UFOs, and I'm reading your book and my motivation is curiosity. Now, what can you tell me about your books compared to what else exists in the literature?

VALLÉE: Okay, I wrote my first book about it [*Anatomy of a Phenomenon*] out of a feeling of responsibility, because I discovered that by studying this phenomenon I had come into a position of knowing better what was the signal and what was the noise. I was in the unusual position of having analyzed the original data both in Europe and in the U.S. Frankly, I thought at the time that the best explanation, the best hypothesis was that we were indeed being visited by a civilization from outer space. That seemed to be the most logical explanation, the most probable. If that was the case, there was a certain probability that something would develop, a form of direct contact that would be important for all of us, as a society; then scientists might have to make decisions fairly rapidly on the basis of a whole series of new sightings.

If they went to the literature, they would have to do the same amount of work that I had been doing in three or four years, and there was no one place where documented cases were gathered, without all the noise and the random sightings and the fakes and the hoaxes. So I felt it was my responsibility to place the best cases on record, even though I did not have the theory that would explain these cases. I felt that the other parts of the puzzle might belong to biologists, physicists, or psychologists who were out there unknown to me. I was in a pretty unusual situation. And that's why I wrote *Anatomy of a Phenomenon* as a survey of what people had said about the sightings and what the best sightings had been.

HASTINGS: This was the book in which you cataloged what the general patterns of sightings were, is that the one? Or was that *Challenge to Science*?

VALLÉE: *Anatomy* only surveyed the different approaches and the spectrum of classifications. *Challenge to Science* cataloged the patterns and was more of a scientific approach to UFOs. I wrote that one pretty carefully as a monograph, and Allen wrote the foreword. The main point was that science already had a lot of techniques that could be applied to the analysis of a phenomenon such as UFOs. I argued that we do not need to phrase it as a belief problem, we can take the whole question out of the issue: "Shouldn't I believe or should I believe?" This is a false problem—you don't have to believe in electrons to study electronics, and you can observe a phenomenon and study it and apply analytic techniques to it without a strong emotional commitment to a particular theory of it.

HYNEK: You don't have to believe in electricity; all you have to do is pay the bill!

VALLÉE: After that second book, there was a gap of a few years during which I was fairly disgusted with science, and I emerged from it with the manuscript of *Passport to Magonia*, which was not a scientific book but attempted to look at the relationship between modern UFO sightings and folklore. I was saying that if UFOs are nothing else, they are very important modern folklore. Now, human imagination has certain laws just as stars have laws and electrons have laws. Human imagination follows certain patterns; it does not behave randomly. I was trying to isolate the laws of imagination as they are related to UFO sightings and especially to landings and close encounters—showing that UFO occupants behaved much like the elves and fairies of old. In that book, I also compiled a catalog of nearly one thousand landings.

HASTINGS: And Allen's book, *The UFO Experience*?

HYNEK: Mine was an attempt to gain some currency for the subject in the scientific community. I picked as my audience the level of intelligence of the people who would read *National Geographic* or something of that sort. I didn't want it to be a popular book and yet I wanted to serve the purpose of people who were on the fence, who had been reading about UFOs and never paying any real attention. My audience was people who may have been saying, "What is all this UFO stuff about anyway? Is there anything to it, or is it all nonsense?" So that's the way I started; I simply said, here is what is being observed, and it seems to fall into certain patterns. There were six categories: nocturnal lights, daylight discs, and so forth. I would take at least a dozen cases in each category that I had some personal acquaintance with and make a model of that particular category using those cases. I also used the book as a vehicle to expose the fallacies in the Air Force and the Condon Committee.

Should We Burn the Condon Report?

HASTINGS: Speaking of the Condon Committee Report, how would you suggest that people look at that book or understand it? If I have a copy of it, what do I make of it? How do I approach it?

VALLÉE: We have a French colleague who suddenly developed an intense interest in the subject; when Dr. Hynek asked him, "How did you become convinced of the reality of the UFOs?" He said, "By reading the Condon Report, because when I saw that someone was taking so much trouble to explain something away, then I realized there must be something to it!"

We heard a suggestion the other day that seems to be very good. Someone said, let us republish the Condon Report *backwards*. The report has Condon's conclusions first, you see, conveying the impression everything is explained, and then a lot of irrelevant material (known as "padding") about how radar works, and then all the cases are found later. Well, a significant proportion of those cases is *unexplained*. What one should do is print the cases first and throw away all the pages about radar. In case after case, the investigators said "a definite UFO here" or "this case cannot be explained other than by an external intervention," and so on . . . then at the end we would print the conclusions of Dr. Condon. I think people would gain a different perspective of his biases.

HYNEK: So the answer to your question of how should the Condon Report be read, the answer is *backwards*. The cases first, that's the only important thing, and virtually draw your own conclusions!

The Amateur Groups

HASTINGS: These private groups, NICAP and APRO. Who are they and what have they done? How would you want to interface or share your data with them?

VALLÉE: The fate of all those groups has been linked to the level of visibility of the overall phenomenon. When the Air Force stopped doing investigations, the amateurs lost lots of members. There were times in '63 and '70 when it was not clear if APRO and NICAP were going to survive.

HYNEK: In '73, they really needed someone to help them, because they needed lots of new members. Of course, their besetting sin is that they exist by popular membership. So anybody who has ten bucks can join. They virtually have no way of screening out crackpots.

VALLÉE: The vicious circle here is (a) they sincerely want to do research; (b) to do research they need money; (c) to get money, since they cannot get a grant, they have to go to the public; (d) to get money from the public they have to advertise; (e) to advertise they spend the money they do have.

So 90 percent of the money they do get goes into public relations, and they turn into a PR organization. There is never enough money to do any research.

HYNEK: Their bulletins can hardly be called scientific documents.

VALLÉE: All these groups go through the same stages. They always start with one individual who *really wants to do something*, who is dedicated, who is hardworking; usually he's a pretty good researcher. He sets up the group because he thinks an organization is going to give him more leverage and more structure, and he feels a need for structure. So he sets up the organization and pretty soon it attracts the kind of people we were talking about earlier, the curiosity seekers, the people on ego-gratification trips.

In the end, they find themselves actually *hiding* information instead of revealing it, thus going exactly contrary to their stated purpose. You see, there are lots of inter-organization feuds. Each organization derives its identity from the files it has. So the whole idea is predicated on keeping those files to themselves and publishing almost nothing.

HYNEK: They're publishing just enough to titillate the interest of their subscribers.

VALLÉE: They keep up their image as investigators, but, of course, that kind of thing is useless to someone who is trying to do real research. So if you buy the typical bulletin of one of those groups, you will have a two-page editorial that says Mr. Such-and-Such has resigned; Mrs. So-and-So's knees aren't getting any better (club news); So-and-So gave a lecture and it was enthusiastically reported in the *Morning Star Register;* and, we're making great progress. That's about it. The people in those groups are sincere and do good work, but it doesn't get out. To a large extent they have saved a lot of information from oblivion. On the other hand, the role they are playing now is not a scientific role but a sociological role.

HASTINGS: What do they need to do? Give them some advice.

VALLÉE: Publish!

HYNEK: Right. They rarely publish documented data. What would one think of a research institute in existence for twenty years that has not turned out one single scientific paper? That's the situation.

HASTINGS: How are they in investigation?

HYNEK: That is fairly good. They have some good people and some tremendously devoted people. I'm thinking for instance of Ray Fowler, in Massachusetts, who was associated with NICAP and turns out beautiful reports. He has recently published a book called *UFOs—Interplanetary Visitors* that tells the inside story of his investigations.

HASTINGS: They don't publish these reports in ways that make them scientifically available.

HYNEK: No.

VALLÉE: Most of the time, they don't even know what they have in their own files.

HASTINGS: Do they have training programs in terms of training people how to investigate?

HYNEK: Well, they say they have.

VALLÉE: They send out a little sheet from time to time.

HASTINGS: Do they have conferences, meetings, conventions?

HYNEK: MUFON has an annual convention and some pretty good papers, and to that extent they are publishing. APRO has had some good conferences too.

VALLÉE: I am disappointed in these organizations, because they create the type of thing we don't really need in this field. Each one sets itself up as an "in" group and puts everybody else "out." The only real group I know that hasn't done that is *Lumieres dans la Nuit* in France. They say, "We are not going to set up an organization, we are not going to centralize, there will be no central files." They just publish everything they get from a number of local "circles." Locally, anybody can start creating such a group or "circle" provided they meet more or less regularly, do real investigations, and send those investigations to the man who publishes them.

The point is that he doesn't tell them what to do. As long as the investigation meets certain standards of sincerity and accuracy, it is published immediately. At a later time, they go back over this information. They have people who specialize in doing physical research, statistics, correlations, and so on, but that's something you do afterwards. You don't sit on the data for ten years until you've had a chance to discover "the truth" and reveal it to the people! The idea that any one of those groups is going to discover "the truth" about UFOs and "scoop" all the others is ridiculous! The problem is too complex.

HASTINGS: Is *Flying Saucer Review* published by a group?

VALLÉE: No, it's not an organized research group. It's a team of essentially three people who are putting the publication together from material they get. They do not have investigators or dues or members or anything like that; they are strictly an editorial team.

HYNEK: People simply subscribe to the magazine.

HASTINGS: It's like a journal then. What's their quality?

VALLÉE: Very high. They have been the only international journal in this field, and they are now in their twentieth year. I don't know where we would be without them.

HYNEK: I heartily agree. I think we should say, too, that the Center for UFO Studies cooperates with any organization or any journal. It should be like that. A colleague of ours has pointed out, and I think wisely, that we should try to publish in recognized scientific journals.

HASTINGS: So what professional journals at present seem open to ...

HYNEK: *Astronautics and Aeronautics*, the AIAA journal, is open-minded. They've published several case investigations, and that's a highly respected journal. And I think we could get things eventually into *Science* and *Nature*.

HASTINGS: Jacques, how much do you know about the Japanese groups?

VALLÉE: I know they have been very interested in skywatches. Different groups in Japan organize trips to the hills where they sit with telescopes and flashlights, trying to attract UFOs. I'm rather skeptical of the value of that.

HASTINGS: That might work, who knows? You really can't predict what's going to be productive.

VALLÉE: Most of these groups are simply not doing serious work.

HASTINGS: In a sense that really does fit into the sociology of the thing ...

HYNEK: Yes, UFOs and society.

HASTINGS: At the same time, you can say now that in the scientific field, more and more scientists from a wide variety of fields are getting to be interested in UFOs.

HYNEK: They do this mostly on their own time, but it's of the same professional quality.

HASTINGS: Actually, listening to you all talk, it seems to me there is far more scientific research going on now than there was five or ten years ago. You do really have more support than you think in the scientific mainstream. I think you would gain, not from saying, "we need funds, and we're in terrible shape," but by pointing out that things are starting up. I think that's one of the faults, incidentally, of the parapsychology people. All they do is moan and groan about not having any funds. And that's all true, but that's not what they should be doing.

HYNEK: The negative thing in that is it sort of creates an additional negativism.

VALLÉE: For a lot of this kind of thing, you don't need money. You don't need a lot of money.

HASTINGS: I don't think we would want to be trapped by the mainstream anyway. There are disadvantages in being caught up in an establishment.

VALLÉE: That's what I keep saying to my friends in the French groups! They say, "What we need is scientific recognition." And I say, "*The last thing you need* is scientific recognition. Now is the time to do research! You have all the freedom you want, you can go where you want. The moment you have scientific recognition in France, then the director of Paris Observatory is going to be put in charge of what you publish and who gets the funds . . . you will have a project monitor in Paris who is going to review everything you do, and you won't be able to go and investigate what you want. That will be the end of research!"

Some Intriguing Stories

HASTINGS: Would you recommend the books by Keyhoe?

VALLÉE: Keyhoe wrote *Flying Saucers Are Real,* among other things.

HYNEK: I wouldn't recommend that myself. When I have a seminar on UFOs, I do not put that on the reading list, but if a person wants a sociological perspective, he should have it, because Keyhoe was very largely fighting city hall. He took as his objective to vilify the Air Force, try to get congressional investigations and prove that the Air Force was covering up. What he should have done, of course, was publish solid technical reports on cases and then the sheer weight of those would eventually have toppled the Air Force. But instead, he kept importuning congressmen, and didn't get anywhere. I would think also, if I were to recommend anything in the popular category, I would choose one of Frank Edwards's books.

VALLÉE: The problem with these books is that their value in terms of information doesn't match their entertainment value. So you have no way of using them. If you want to use any one of the cases they quote, you physically have to do all the work all over again. You have to start from the beginning, and of course they don't give you the information you need to start from the beginning!

Now scientists when they publish data *always assume they may be wrong.* A paper submitted for publication is not acceptable if the data, and the way the data have been acquired, and the source of the data are not referenced. It just would not be published. So with books like that, we find ourselves confronted with a lot of observational information which is interesting; anecdotal information which is very valuable or would be valuable if the source was given, but most of the time the source is not given and essential data are missing. These people are good writers in the journalistic sense; they can generate suspense in the reader. They give you the exact information on when they got the story, but they don't tell you what the story was! For example, at 3:18, exactly, the author was just jumping out of the cab about to go into his house when the front page of the *Washington Post* caught his eye, and the headline was "UFO Seen in South America." Well, he may never get around to telling you what happened in South America, but you sure know it was at 3:18 and he was standing exactly at M and 18th NW, and that's all the information you have. It looks very accurate, but when you look at it, there is really nothing you can use in a scientific sense.

HYNEK: None of those books constitute scientific data.

HASTINGS: So you really can't use them as a basis for any kind of understanding.

HYNEK: No, you can't.

HASTINGS: Among the books which are on the fringe areas, such as the Adamski books, what about material from people who claim to have ridden in UFOs and so on?

VALLÉE: There is one intriguing story as far as I'm concerned, in the book by Truman Bethurum called *Aboard a Flying Saucer*. It's not a classical contactee book in the sense that he has not tried to capitalize on his observations, and he is very much puzzled by it. It's not unlike the Betty and Barney Hill case. He says he was in the desert, working on a road-building project, and he found himself isolated in his truck and fell asleep. He woke up with these strange people around him and a large ship in the desert. He says he was taken inside and met the captain of that ship. Now this has some elements of a dream again: The captain of the ship is a beautiful woman, and he has long discussions with her (and apparently it never got more personally involved than that). The case is certainly charming and colorful. Well, that's one in that category.

Another one that must be mentioned as one of the most excellent books that is relevant, especially in view of what's happening now with UFOs, is *When Prophecy Fails*. It is a study by a team of sociologists who infiltrated a UFO sect predicated on the idea that there was going to be a flood in the United States and that the flying saucers were going to save the believers. The authors followed the development of that group and their question was *"What happens to a sect like this when the prophecy fails?"* The answer is: The belief structure becomes reinforced by the failure of a prophecy.

HYNEK: That's a fantastic thing. Another one, Dave Jacobs's dissertation. It has now been published by the Indiana University Press. The title is *The UFO Controversy in America*. That's a must.

HASTINGS: Is that a compendium?

HYNEK: It's a doctoral dissertation, from the history department of the University of Wisconsin.

HASTINGS: *When Prophecy Fails*, by Leon Festinger, Henry Riecken, and Stanley Shackler is published by HarperTorch, Harper, edition 1964. What about the *Interrupted Journey*?

HYNEK: That's the one I was trying to think of. There's another one, however, *Incident at Exeter*. These two are by John Fuller. He does, I think, a very good job of reporting in these two books.

HASTINGS: From your experience, do you find a correspondence between your investigation and his reporting of the case?

HYNEK: Yes. Those two books I would rate very high.

HASTINGS: More recently there is the book by Ralph Blum, *Beyond Earth*, a 1973 Bantam paperback.

VALLÉE: It contains a good account of the Pascagoula incident. Then there is the book by Dave Saunders. It's another classic. Really excellent.

HYNEK: Saunders's book called *UFOs? Yes!*

VALLÉE: It's the inside dope on the Condon Committee. And then, at the other end of the spectrum now, there is this book by a friend of the contactees, *The Stranger at the Pentagon* by Dr. Stranges. This has, among other things, a picture of the Martian that was captured by two FBI agents! In fact it was a monkey who had been hit by a car on the road.

HASTINGS: What ever happened to Otis T. Carr?

VALLÉE: He flew away, didn't he? Got to the moon in his flying saucer?

HASTINGS: Okay, what else is there in terms of books and journals? What about the *Flying Saucer Review*? Would you recommend that?

VALLÉE: It's the only publication in this field of a professional caliber. I think the *Flying Saucer Review* is all you need to know about UFO magazines, at least in the English language.

HYNEK: I would certainly recommend *FSR*, as it is usually referred to. Of course, you might put in Menzel's book as an antidote. [Laughter.]

The Skeptics and the Damned

HASTINGS: What are the really major skeptical books? Menzel's is one.

HYNEK: Menzel's certainly is.

HASTINGS: And he explains everything as temperature inversions and reflections and mirages?

HYNEK: Yes, *everything*. [Laughter.]
   And Phil Klass explains everything as hoax or delusion.

HASTINGS: How does he take care of UFO droppings?

HYNEK: He doesn't, he really just avoids them, or he says they are all hoaxes.... But there are just a few things he does not know about!

VALLÉE: There are a couple of significant books by Jim and Coral Lorenzen. One of them is called *The Shadow of the Unknown*. Also there is an intriguing book, *Operation Trojan Horse*, by John Keel. And now, *The Mothman Prophecies*.

HASTINGS: What about Charles Fort? Of course, it's not scientific documentation, so how would you see those books as being interesting?

HYNEK: I wouldn't put them in myself, except for the sake of historical perspective.

HASTINGS: Just for their historical value . . . ?

HYNEK: Well, the point is that Fort has three chips on his shoulder. He was out to absolutely vilify anyone who looked like a scientist. Scientists were stupid, in his view. The man has completely disregarded the stupendous and positive things that scientists have done. Good Lord, look what's been done in astronomy and celestial mechanics and bacteriology! He made a specialty of picking cockeyed things, not particularly documented, completely uncritically, he just wanted to show up these damn fool scientists.

VALLÉE: Well, I think scientists need people like that to keep them humble! *The Book of the Damned* is a remarkable document, you know.

HYNEK: He'd have been much more effective if he hadn't been quite so blatant, if he'd been a little more scholarly in his presentation. A guy on a soapbox waving his arms at sixteen rpms doesn't inspire confidence.

HASTINGS: That could be said.

HYNEK: But he's got a valuable collection of things there from the standpoint of popular accounts.

VALLÉE: Even from the standpoint of scientific journals. And I like the way he writes.

HYNEK: This recent book on "Strange Phenomena," did you ever get a copy of that? That's another good one. . . .

VALLÉE: Yeah, but it doesn't say anything about UFOs.

HYNEK: Neither does Charles Fort, a great deal. All sorts of fishes that fall from the sky. I hope I'm not being unfair and have not overlooked something, but most of the books just get you nauseated....

HASTINGS: And what about von Däniken?

HYNEK: My personal theory about von Däniken is that he has touched a sensitive nerve in our collective unconscious, in our racial history. But his work is illogical and unscholarly.

HASTINGS: And he's drawn on so many different sources! He's done something similar to what Jung has done, but Jung has done it at such an academic level!

HYNEK: Well, Jung was very puzzled. He wasn't ... he was a puzzled guy, something like Ruppelt.

The Function of Myth

VALLÉE: There is a big gap, there's no sense hiding it. There's a big credibility gap between the scientist and the public in this respect. von Däniken just dramatized the existence of that gap.

HYNEK: He capitalized on calling the professional archeologists dumbbells.

HASTINGS: I'm not so sure he called them dumbbells as much as he went beyond them. It's not just what they say, it's far more. He said more crudely what Gerald Hawkins said in *Stonehenge Decoded;* Hawkins said essentially, archeologists and anthropologists have had the blinders on, and they haven't utilized other sciences to help them understand what Stonehenge is about. And here's another scientist brought in, different perspective and everything else, with the understanding of how computers can help him work out his hypothesis. He indicated at the popular level that science, because of its highly compartmental approach to a problem, can get off on a track and not expose the problem. von Däniken allows the UFO problem to be exposed to more minds and more diverse viewpoints.

VALLÉE: More than that, releasing all this passion and energy has dramatized the fact that there was an immense amount of emotion stored in the UFO problem. But notice that he never talked about the UFO problem!

HASTINGS: One of the functions of myths is to mobilize energy in the society, and when myths fall out of favor, it's because they no longer mobilize energy. You see modern mythology developing out of processes that mobilize energy. And von Däniken and the others, what they're doing in effect is *revising our past;* our present needs certain kinds of changes and assumptions that revising the past will provide. The technique is similar to psychoanalysis, in which you revise your own past so as to change certain kinds of habits and patterns. You change, in effect, how you evaluated your reinforcement in the past. And I see one of the things that von Däniken is doing as revising our past so as to give us certain different perspectives of ourselves and our place in the universe. Obviously, it's capturing a great deal of energy, which means, to my point of view, that people want to accept those kinds of changed assumptions, and this is a convenient way to do it. And they may very well be true, but that's not why it mobilizes the energy. Lots of things are true but don't mobilize energy.

VALLÉE: But it's very dangerous to do that. It's a little bit like Hitler rewriting the history of the Aryan race.

HYNEK: Or Russia rewriting Tsarist history.

HASTINGS: Well, that's a different question. That is to say, given that it is happening, given that point of view, what do we make of it? Is it dangerous? What does it imply?

VALLÉE: It's healthy in the sense that it exposes contradictions, in existing anthropology and in existing ideas we have of our culture, where there is a lot of material that has been swept under the rug.

HASTINGS: I have a paperback called *The Flying Saucer Reader* which has one of your chapters in it, Jacques, plus a variety of other things. . . .

VALLÉE: *The Humanoids* by Charles Bowen is a good compilation that is in print. It is a good collection of articles on the subject of occupants, something like three hundred cases, documented, summarized. . . .

HYNEK: I'd include that, and then I think one more book is *The Age of Flying Saucers* by Paris Flammonde. It is amusing.

HASTINGS: Is there anything on the psychological aspects?

VALLÉE: No, psychologists take themselves too seriously to approach the subject.

9
Brainstorming

Hypotheses and Other Universes

Some Scenarios

For their final session, the authors and Dr. Hastings decided to discuss what they see as the possibilities in the years ahead, what might be the future behavior of UFOs and the developments of UFO problems. Author Vallée began by asking, "What if we were writing 'The UFO in the Year 2000'?" He continued:

VALLÉE: Let's brainstorm about that. . . . We would have various scenarios.

Scenario No. 1: Things keep happening as they have for twenty-five years, and we keep publishing more books about it. Blum & Blum publish more books about it, von Däniken publishes more books, and nothing else happens! Every two or three years there's a flap somewhere. There is no visible effect on society, there is no direct threat, there is no mass landing; that's one scenario. And we can talk about the consequences of that, what that could mean. Certainly, some theories about it would have to change. Right?

HYNEK: Well, the theory that they are going to invade us would have to change!

VALLÉE: Yes. If they are hostile, and they've been hostile for twenty-five years, they're certainly not doing a very good job of invading us! Then there may be another scenario, which is at the other end.

Scenario No. 2: They land, they take over. Okay. They say, "We're from Zeta Reticuli, and this business has been going on too long!" I call this the "Twilight Bar" scenario, after the play by Arthur Koestler.

HYNEK: They say, "We need more protein, and you're protein!"

VALLÉE: Or, "We don't want you to make H-bombs anymore, and you have to sign this treaty which every planet signs to be accepted into the galactic federation." That's one other thing.

Scenario No. 3: They could stop and go away entirely.

HYNEK: That was my fear, you know, when the Condon committee started. I thought, Oh boy, this is the way things go. It's like going to the dentist with a toothache! The toothache disappears when you get there. I thought, well, as soon as the Condon committee gets going, wouldn't it be ironic if UFO sightings ceased! You know, things work out like that sometimes—a genuine perversity of nature.

VALLÉE: Well, of course, it was almost that—there were very few cases.

HYNEK: Yes, it was that.

VALLÉE: There was a very big wave, but it was in Spain, and, of course, Condon didn't know anything about it. He didn't care. I asked Condon if his project would look at the international scene, and he said, "No, we are only looking into the American sightings!"

HASTINGS: It's like looking for meteors just over the United States!

VALLÉE: What are the other possible scenarios?

HASTINGS: Okay, here is another one.

Scenario No. 4: We learn to contact them. We set up a crash program to communicate with them telepathically, using all the characteristics we have learned that somehow seem to be associated with landings and, suddenly, there's a wave. So we send out people who are trained to watch the skies and to hold up signs, or we send people driving down roads, hoping they'll see some broken-down spacecraft on the road. We attempt to contact whatever it may be physically, telepathically.

HYNEK: We set up a hotline for everybody, instead of just the police.

HASTINGS: Or train people to go into astral projection, and every time UFO waves come, we send those people to try to use clairvoyance to tune in on them. So there's another scenario. We learn to contact them, and we discover that they are subjective rather than objective, though they have certain abilities to generate physical phenomena as well as subjective phenomena.

VALLÉE: What's the effect on our civilization? What is the impact?

HASTINGS: An elite develops. The people who can contact them versus the people who can't, and the elite have control over the rest of society.

VALLÉE: So we come into a neo-Egyptian type of hierarchy. The pharaoh is surrounded by high priests who have contacted higher entities. The people are reduced to a slavish level.

HASTINGS: And so, actually, they arise and burn them all. Society breaks down. A new witch hunt arises, and in fact, there is a scenario of that sort in Clifford Simak's book, *Out of Their Minds*, in which we are exploring planets by using a machine which enables us to mentally bring ideas back, and the company making this machine becomes a conglomerate to end all conglomerates—controls everything. Simak, incidentally, do you know who he is? A Minneapolis newspaper writer who has written at least two dozen stories about trafficking with aliens over the last twenty years. He has explored many of the possible implications of how you communicate with them and what impact it will have. I have toyed with the fantasy at times that he knows, that he is programmed to do this! Because he really is accurate. He just knows how to make it sound pretty real. He's written about travel to other space, other dimensions. The whole range of things. There's another scenario.

Scenario No. 5: The Air Force shoots one down and they attack, they block all electrical communications.

HYNEK: "Let's shoot one down and see if they're friendly."

VALLÉE: There was a very good survey done in Detroit about UFOs. They asked people what we should do about them. Twenty-six percent of the people said we should shoot them down, and 74 percent said we shouldn't. Then they asked more detailed questions. Someone said, "We ought to shoot them down because it's the American way!"

HASTINGS: It's the American way to shoot them down?

VALLÉE: Yes. And one man said, "We shouldn't shoot them down, they couldn't possibly be weirder than some of the things that walk around the streets in Detroit!" And another one said, "We've got enough people on welfare without supporting a bunch of Martians, too," so we should shoot them down because of that.[1]

The Billionaire Friend

HASTINGS: Okay, here's a scenario and it's somewhat plausible.

Scenario No. 6: Let's say you have a billionaire friend who is interested in helping out, and you say to him, look, a lot of people are seeing them but they're

---

[1] Telephone poll conducted by the Detroit *Free Press*, August 1971.

not reporting them. What I want to do is set up this universal hotline service. Okay, so we now have twice as many or five times as many reports, or even twenty times as many, and the reports are about the same; again, not much evidence but the same quality.

VALLÉE: They would be of the same general quality, that's one point we can reinforce. All the statistics are about the same. One thing that comes out of the French study is that if you take the best two hundred or three hundred French cases (those are not from the public, they are highly investigated cases, and the full treatment on each one), you get the same thing as when you take the refined U.S. Air Force data; it's all very stable. No matter what sample you take, it's practically the same thing; you just reinforce the pattern. All the data are good, virtually, once you have eliminated the really obvious errors, the balloons and the mirages.

HASTINGS: Then what do we want any new data for?

VALLÉE: It isn't clear to me that we need a lot of new data except to confirm some patterns.

HASTINGS: Wouldn't you try to set up situations then, if you had your funds? You'd plot your cycle and your locations from the beginning of a cycle, and you would move out to the location and expect to see one.

HYNEK: Sooner or later, we're going to have a predictive possibility like this. But the reason we have our police hotline now is not merely to get more data. We couldn't care less how many more lights in the night sky we get, but we want those few cases, like this Ely, Nevada, observation—those are the ones we want to go after. This is the scenario I would like to see happen: a serious study program gets underway and leads to the discovery of new scientific facts, new paradigms.

HASTINGS: We're talking about scenarios—say you had the funds and you were able to predict the time, location, etc., because of the wave pattern building up—how would you want to interfere? You want to be there. You want to attract their attention, you want them to understand that you want to talk. Get out of this damn thing and let's talk and let's find out what's going on! You're ready for it, you can handle anything they've got. How do you go about that?

HYNEK: Well, if their sailing orders are not, under any circumstances, to make contact, there isn't anything we could do about it.

HASTINGS: But they have made contact, we suppose, but it's been a one-way sort of thing.

HYNEK: Yes, it's been completely a sort of exploratory maneuver.

VALLÉE: But it's been intentional.

HASTINGS: It appears to be. Even by our own standards of behavior, there's some kind of intent. Now, we're all describing it as if it were outside of ourselves. We are assuming there's no psychic thing going on; but disregarding that, wouldn't you attempt to set up some experiments to see if you could make contact?

HYNEK: Keyhoe, in his book, has suggested "Project Lure," setting up a dummy UFO—a UFO decoy.

HASTINGS: You mean like the cargo cult? Like the aborigines of New Guinea setting up dummy airports in the jungle to attract airplanes?

HYNEK: Yes, some of the aspects of cargo cult. But it would probably just elicit a smile, you know, "Gee! These earth people are really stupid."

VALLÉE: One way to make me run away if I were an alien visitor is to see those fellows sitting there in the desert waiting for me! Another scenario—

Scenario No. 7: Somebody publishes a book that explains what they are—somebody discovers what they are. Now they have to do something about that! Somebody exposes them, explains the whole thing.... It would be interesting to find that the behavior of UFOs changes after certain books are published. For example, perhaps sightings did occur along straight lines until Aimé Michel wrote *The Flying Saucers and the Straight-Line Mystery*!

Four Hypotheses

VALLÉE: Let's try to develop some basic hypotheses. Let's map out this space.

HYNEK: We need to line up all the things that a viable theory needs to explain; we don't have anything in the present framework of science that would explain it. How would we have to alter our present scientific belief scheme—what theories, what hypotheses? Is there a hypothesis, no matter how strange, that explains the facts?

VALLÉE: Within the assumption that UFOs are alien, there are two categories I can see where we can start developing hypotheses. They could be either earthbound or extraterrestrial aliens. Then, there are two other categories, where the thing *is not alien at all.* Maybe there's someone in a cave who is remotely dispatching material objects in Kansas or California, Missouri and France? Maybe those are holograms that some clever scientists are projecting? Maybe there is a highly instrumented place somewhere from which people are emitting all those things? And another series of hypotheses, where the UFO phenomenon is purely human, is the framework where we assume that it's genetically programmed. In this framework, it has nothing to do with any external object. It may be that there is inherent in our species, a sort of built-in defense mechanism that reveals itself only in times of extreme social stress, and that one of its manifestations is the phenomenon known as UFOs.

Remember, there has never been such social pressures as we have now, and such changes in the environment... the time is unique! There have never been so many people on the earth, so many pressures on them, there have never been so many dangers. That certainly produces some very unusual biological consequences. There might be a triggering of a genetic process that has never happened before. That might be manifested, among other things, by people seeing things in the sky, and various physical effects might even result.

HYNEK: Programming would be producing the vision of things in the sky?

VALLÉE: Genetic programming. It is only one of four categories of hypotheses. Two of them involve an alien intervention, and the other two a human source. To summarize: Hypothesis 1A is *outside alien*, and 1B is *earthbound alien*. Hypothesis 2A is a *secret human base*, and hypothesis 2B is *genetic programming*. One of them, namely the last one, doesn't assume any kind of technology. It's just a normal, natural process; all the others assume a novel technology. Now we can talk about specific ideas within these hypotheses.

HYNEK: Is it a logical way to present it?

HASTINGS: I can see one advantage in that it opens up the minds of people to things they wouldn't normally think of. It does not limit us to assuming that it is just visitors from outer space. Is it people from the earth itself, and so on? We bring them into consciousness, so to speak, and we say, look, there's evidence for and against each of these, and we don't have to accept all or any.

VALLÉE: Do you have my hypotheses? Let's start with 1A, which is outside alien forms—here you have the flying saucers from outer space, gods from outer space, Martians, Venusians, space brothers, galactic civilizations, and so on. What do we have to say on that?

HYNEK: The whole concept of life elsewhere... the idea that we are being visited. There are two concepts I would like to bring in here for the record: one is how insignificant the earth actually is, how utterly improbable it is that we would be the only inhabited planet. It is highly improbable that we would be the only ones. Chances against it are trillions and trillions. On a model where the visible universe would be the size of the U.S., the earth would be invisible even with an electron microscope! That raises the question that Carl Sagan comes up with every once in a while, "Well, since we are so tiny, why would anyone bother with us?" What are the probabilities that, even if there are all sorts of civilizations, we would ever be visited?

VALLÉE: That is one liability of the hypothesis that we're dealing with space visitors.

HYNEK: That is one liability, yes.

VALLÉE: Another liability is that they wouldn't do it this way; at least, we wouldn't expect them to do it this way. First, the investment this would represent with our ideas of a scientific mission would be out of scale. It wouldn't make sense to send thousands of spaceships into Kansas, where there is practically nobody around, to pick up some branches again and again and again for twenty-five years. Are there any things that we can say to counter this kind of argument?

HYNEK: There is one thing we can say against that projection of "Why us?" I think it should be pointed out that the human race has been here for a few million years and, during that time, what's it done? For four million minus a hundred, it has done nothing except fight wars. This is an epoch, a time in history when science has suddenly appeared.

VALLÉE: So? For the last hundred years we have done nothing but fight wars too!

HYNEK: Well, we've gotten from Kitty Hawk to the moon in seventy years.

VALLÉE: The only difference is that we are fighting bigger and bigger wars.

HYNEK: All right, but I would still like to play this hypothesis of Lunan's.[2] Let us suppose that NASA in a million years had gotten powerful enough to start investigating other solar systems and a quick survey shows that there is no intelligent life there at all. But it sets up an artificial planet in the system as the cosmic monitor that monitors every once in a while, and maybe once in every one thousand years reports back to the home base. Then suppose NASA has lots of probes in various parts of the universe. It has ten thousand of them out. It is like looking for a supernova. In any one galaxy, a supernova occurs only once in two hundred years, but by searching thousands of galaxies, we catch several supernovae a year. So, suddenly, after a million or so years, the probe in this solar system starts reporting back, "Aha, the human race, this little protoplasm that has been developing here, is suddenly getting interesting and bears a little more watching."

HASTINGS: Particularly if they have picked up radiation!

HYNEK: Radiation, yes; so that I think the objection can be knocked down by saying that the human race may have become cosmically interesting. Now, also, I would like to play around with a scenario which is somewhat on a different track; let us suppose that in its explorations, NASA comes across a planet on which

---

[2] Duncan Lunan, a Scottish astronomer.

there's a civilization which has arrived at the level we were, say at the year 800 or 900; what would our sailing instructions be? What would NASA instruct its astronauts to do or not do?

It's possible they would instruct them not to interfere with the natives, for fear of creating a massive cultural shock. They also might instruct them to confuse as much as possible, to do stupid things so that the people would not be believed if they saw something. "It is no spacecraft ... it's ridiculous! Why, they are stopping cars, frightening animals, and doing all sorts of crazy things; no sensible people would come to do something like that." In other words, a counter-espionage technique.

VALLÉE: But if you are going to do that, you could accomplish the same thing without landing on the planet. You could do that remotely.

HYNEK: All right, that's much better. Because NASA in a million years is not going to be sending large space probes, they will have miniaturization ... in the same way that you couldn't compare electronic tubes to the circuits of modern computers. So do we know what the miniaturization of a million years from now will be like? I would think the visiting robots or life form (or maybe the distinction between robot and life form may become less and less) would be instructed not to interfere with that particular scheme of evolution; we have no moral right to interfere, but we are curious and we want to find out.

VALLÉE: That's an argument against UFOs being alien space probes, then. Because, of course, you would do that by automatic means without being detected; just as there are at least two nations on earth studying the whole planet right now, square mile by square mile, without being detected. You can see stripes on parking lots. ... You can even see through haze, in the infrared range. We can do all that now. Do you realize that some of the radio communications we have today could not have been detected with the radio equipment we had ten years ago? The best radio equipment available ten years ago couldn't detect the messages that are now transmitted by radio waves in computer-to-computer communication, both cross-country communication and trans-oceanic communication. So what does it mean in terms of the speed at which our technology has evolved?

When Sagan and others come along and say that we should try to detect radio communication from advanced civilizations, they are assuming that civilizations a thousand years ahead of us, or three thousand years ahead of us or millions of years ahead of us, are using radio communication that we can detect. Is that reasonable? The chances that they would still be using such an obsolete mode of communicating are very small. I think that it is something that we should really emphasize ... the most advanced type of radio communications we now have could not have been detected by the best equipment ten years ago.

HYNEK: Let's come back to that later. If we don't have an answer to the problem, and we don't, we still can do something positive for people and really give them something to think about.

The Nature of Contact

HASTINGS: How about the following hypothesis? If it is not monitoring, because monitoring could be done more efficiently by other means, maybe it is a deliberate attempt to influence us by doing things that will call attention to anomalies ... non-realities.

HYNEK: Now that ... I buy that. Jacques, you used some sort of phrase the other day ... you used some phrase that escapes me now, a *conditioning*? Can you expand on that a little bit?

VALLÉE: I said I thought it was a *control system*, a way to condition our social behavior.[3]

HYNEK: Well, suppose that NASA a million years from now decides that it can do something for a culture that is taking the wrong track, and it is evident that it is bound for its own self-destruction; it decides out of the kindness of its heart that NASA should appropriate some money to change the culture. They couldn't come in and take over, but they could better change the culture from within. They still have a hundred years or two hundred years to do it. How would they go about changing that culture?

VALLÉE: One thing that we should bring up here is what Aimé Michel is saying about the nature of contact.[4] Is it necessarily true that we should be able to understand the purpose of the contact? Does a cat understand his master when he comes back from work? When I come back from work, I sit at that computer terminal on the table over there and talk to some guy in London about things that are not even going to exist for another ten years!

HYNEK: I like this analogy.

VALLÉE: Suppose I have a dog and the dog sees that; it doesn't make any sense in the dog world. Yet, we are sharing the same room, the same house, the same space. I say hello to the dog, and the dog responds to me.

---

[3] Vallée has developed this theme in his book, *The Invisible College* (New York: E. P. Dutton, 1976).

[4] See in particular the article by Aimé Michel in C. Bowen, *The Humanoids* (Henry Regnery Company, 1974).

HASTINGS: I sometimes see small bugs that get into our house, and I don't care to kill bugs; so I pick them up very carefully and put them outside. They probably don't know what is going on. All they know is that some strange force has picked them up and waved them through the air and suddenly they find themselves outside. Now, for them, that is very traumatic. But if they had stayed in the house, they would probably have gotten squashed.

VALLÉE: It's not traumatic because their attention span is so limited. It's traumatic for a few seconds.

HASTINGS: So you're saying, of course, that we can speculate on the motives that we might have if we were visiting someone, but that doesn't mean that an outside race that has known us would have those same motives or that we would even have common grounds....

VALLÉE: What about *earthbound aliens* now for a hypothesis?

HYNEK: All right, that's a good topic, it's a good expression, *earthbound aliens*. . . .

VALLÉE: Because one of the things that doesn't make any sense, one of the main reasons that the "control system" has to be labeled speculation, is that there is evidence, as you know, that this phenomenon has always existed and that we are only dealing with a fairly late form of it.

HYNEK: Certainly as early as 1897, and even much earlier.

VALLÉE: We have good evidence now that every culture on earth has a tradition about little people, endowed with supernormal power, doing things like flying, kidnapping people, taking them away, sometimes cohabiting with human females, sometimes being useful to humans, sometimes being playful, sometimes being destructive, very unpredictable; these traditions are not just limited to Ireland or to Brittany, they are all over the world. They are with the American Indians, the Eskimos, the Chinese, the Japanese, the Russians, everywhere, every culture.

HASTINGS: I think that it's important to specify that there is a tradition like this in every culture. In connection with these, do they have vehicles?

VALLÉE: In American Indian folklore, people describe "baskets that come down from the sky."

HYNEK: As I mentioned before, the Sioux Indians say that the sky people, when they left, turned themselves into arrows and went up.

VALLÉE: Okay, so what's wrong with that as an explanation for UFOs? Well, for one thing we would have expected to find artifacts. Geologists and archeologists would find them all over the place. Yet, we haven't found an engine or a computer or a helmet or anything like that.

HASTINGS: With an earthbound alien, you wouldn't expect to find computers or helmets or anything.

VALLÉE: Why not? Their civilization would have been going through stages too. There is a French tradition that there were "Lutins" in Brittany until about 1850, and then the steam engine and the Industrial Revolution forced the little creatures to go away, and they disappeared. That's what the old folks say in France. There is the same tradition in Great Britain and in other parts of the world. They went away as civilization went into the industrial age. Well, where did they go away to? You expect to go into their caves and find some artifacts, and you don't.

Stretching the Imagination

HASTINGS: Suppose we discover that there is a way to materialize and dematerialize things; suddenly, we wouldn't leave artifacts anymore, we wouldn't leave any material things behind.

HYNEK: The things self-destruct in ten years.

VALLÉE: Sure, but still our houses and the old buildings would stay behind. And the traces of it on the landscape.

HASTINGS: Arthur C. Clarke in *The City and the Stars* postulates a society in which everything is basically produced in thought forms by a computer. You think of what you want, and the computer goes into its circuits and produces it, projects it. It's three dimensional and solid.

HYNEK: The day might come in which NASA might no longer just send nuts and bolts hardware to Mars and Jupiter but send a thought form that will then materialize there and beam down. That's as strange to us as nuclear energy would have been to Benjamin Franklin, but someday it may be perfectly normal. As Sir James Jeans said, "It's the unexpected that happens in science," and there's no way of predicting what science is going to know fifty years from now.

In 1950, at the half-century mark, the Franklin Institute wanted to query a whole bunch of scientists to see what science would be doing in the next half-century, the latter half of the twentieth century. Well, in astronomy, they asked Otto Struve, who was one of the ranking astronomers of that time, to query fifty astronomers and ask them to predict what they thought would happen in the next fifty years. That was 1950. It was published in 1951. Not *one* of those fifty astronomers said anything about the space age or being on the moon or out in space, or anything like that. *And just six years later it happened!* Now, when you have something like that . . . how can they even think. . . . However, we have to try, and one of the possibilities is projecting thought forms, but communication may finally take a form that we can't even imagine. I'd like to stretch our imagination field by saying that there are some things that none of us can imagine. Could the guy going across the plains states in his covered wagon, could he have even imagined the H-bomb?

HASTINGS: Or a Boeing 747 streaking overhead? How could he have accounted for it if he heard the thing flying over? Or getting from coast to coast in five hours?

HYNEK: Yes, well, he might have imagined that, but his concept of greater speed would have been speedier horses or more streamlined covered wagons; he couldn't have thought of a jet age. I sort of think those things are imagination stretchers.

HASTINGS: To let our imaginations rise! . . .

VALLÉE: Yes, but then, see, people ask me when I give a lecture, "What are your hypotheses? Give us some hypotheses!" I always say that hypotheses are a dime a dozen. I could stand here and give you fifty hypotheses in the next half hour. So what? That's not how science works! A scientist doesn't produce a hypothesis unless he produces with it a way of checking it, and I think we should clarify that: I could make the hypothesis that sunspots are caused by a giant butterfly; all that proves is that I have a wild imagination. Anybody can drop acid and get some very strange ideas like this, and some people think there is value in that, but there isn't. A hypothesis is cheap unless you come up with a statement of how you will check it. If I say, here are three things that sunspots could be and here is how you go about checking them; now, I'm doing something scientific. But this concept that scientists are people who dream up hypotheses is really childish.

HYNEK: The thing brings us again back to Benjamin Franklin. Suppose he was driven to the wall with a stiletto up against his Adam's apple, "Damn it, Ben, what do you think makes the sun shine?" Just like when people come to you after a lecture, "Now what do you really believe a UFO is?" What could he have said? He could have said, "Well, the angels make the sun shine." Or, in fear of his life, what other things could he have said? "It's the work of the devil!" Would he ever have said it's nuclear energy?

HASTINGS: He could have said it's a form of energy that isn't chemical. Because, if it were, the sun would be doing this and that. So we can say that about UFOs. We can say, "Look, it's a form of consciousness that isn't any one of those we know about; otherwise it would behave like this, and it doesn't."

VALLÉE: People ask, "How come they are hostile in some cases?"

HYNEK: Well, if you stick your finger in an electric socket, does that mean that the electric company is hostile?

The Evolution of Man

HASTINGS: I'm still stuck on the alien thing. Obviously, we haven't anything to say pro and con on it. A lot of contact cases and close encounters show creatures that conform with earlier mythological and anthropological descriptions.

VALLÉE: Yes, but it doesn't agree with biology. That's another thing that should be pointed out. Some researchers are getting around this by saying that the witnesses must be wrong about the creatures that do not fit—the monsters, the tiny creatures, and so on. They want to concentrate on the "real ones," the little dwarves with oversized heads that they believe are consistent with biology in the sense of advanced evolution. The idea that intellectual differences are due to the size and weight of the brain is silly. Asian races have a larger brain than the White race. The White race has a statistically larger brain than the Black race, but there is no significant psychological difference. We know dumb Chinese and we know bright Black people, and there is no real evidence pointing to an evolution of the weight of the brain or size of the head as related to intelligence. There is no real evidence that the size of the brain is still increasing and that the next step of the evolution is somebody with a large head.

HASTINGS: But it's not biologically impossible.

VALLÉE: Well, it's biologically impossible if you take into account the whole range of forms and shapes that are reported for the occupants. Unless you keep one type and reject all the others; there is some consistency, but look at the claw men of Mississippi. What kind of creature were they?

HASTINGS: It's probable that a man more than twice normal size cannot survive biologically, simply because the biological process won't support him, but we do have midgets and dwarves.

HYNEK: Not much below that because you get the ratio of surface to mass to a critical value. Three feet is probably our limit, for adult humans.

VALLÉE: We should question some standard ideas about evolution. To the same extent that we should explode some myths about physics, we should explode some myths about biology.

HYNEK: There was one biologist who pointed out that we don't have biologies, we have biology; we have only *one* thing to compare with and that is biology on earth. There may be other biologies, but we know only one, and perhaps we think that everything has to be that one.

VALLÉE: Okay, so on the earthbound alien hypothesis, what do we want to say? Do we want to say that it's not consistent with biology as we know it?

But maybe that's not too surprising. What is surprising is that we don't have more artifacts.

HASTINGS: You would expect to have many more descriptions, and you would expect to have a lot more contacts?

VALLÉE: What we seem to find in archeology is that there is some evidence for contact with something else, but not on a long-term basis. And there we are. Okay, what about the thing being human? Could it be a *human* phenomenon? In other words, *do we really need aliens to explain UFOs if they are real?* Or could the human race have been developed in a very remote past out of a contact between an advanced race of primates and extraterrestrial visitors?

HASTINGS: Okay, many of the creatures are humanoid in appearance, that's one of the theories. To say they are human is more likely than if they came from some other planet.

VALLÉE: And also, the expression on their faces in close-encounter cases. Picture a witness like Mr. Masse in France, scared to death, a few feet away from a little guy with a face twice as big as a human face, who just came out of a flying saucer; he is looking at him and there is a tremendous exchange of nonverbal communication. There are *feelings of communication as there would be with a human being*. For example, Masse said, "They talked among themselves; they were looking at me with amusement on their faces, but they were gentle and wise. . . ." There was this feeling conveyed, of concern and mild amusement. But when you look at a whale, when you look at a porpoise, when you look at a bird, you can't tell what they are feeling! Now if there is real communication, it would mean that the thing is human, wouldn't it? How can they breathe our air? How come there are men, ordinary men, with them? There are cases in France where the witnesses have seen two dwarves coming out of a landed object, and one man with them.

HYNEK: That's the important thing right there, and they don't have to have all kinds of fancy apparatus. They also seem to adjust to our gravity. They don't even seem to wobble too much like the astronauts on the moon. You could tell that the astronauts were not familiar with that environment up there. But these damn things adjust pretty well to our planet. This is an important point that is often overlooked. The chances that conditions on a "home planet" would be almost exactly like those on earth are pretty small. Beings from such places would, it seems, have to wear special apparatus to be able to breathe in our atmosphere, and unless their home gravity was very much like ours, they would find walking difficult. Of course, if these reported creatures are really robots, that would solve this problem.

Secret Weapons

VALLÉE: Well, Dr. Hynek, suppose I play the hard-nosed journalist again: Could it be that the U.S. Air Force is making those flying saucers in some classified laboratory underground and keeping it a secret?

HYNEK: Hardly!

VALLÉE: Maybe there's somebody around who has discovered some very advanced principle. What's wrong with that? That's a much less expensive hypothesis than some strange civilization on Alpha Centauri.

HYNEK: All right. I think we have to examine that. When I answer that question in lectures, I shoot that explanation down with several arguments. One, as I say, it would be difficult to keep these developments a secret that long, for twenty-five years. Some Jack Anderson would have come out with the whole thing!

VALLÉE: Do you really think that there are no secrets?

HYNEK: There are, but not of that magnitude. I would think that, certainly, this technique you were referring to earlier, this undetectable radio device—a few years from now that's going to be common knowledge.

VALLÉE: But do you think anybody will really ever know why there was a gas shortage in 1974? How was it done? The government has thousands of people trying to find an explanation for that, and they can't! People have all sorts of mental blocks that can be exploited if you want to keep something secret.

HYNEK: I don't think that's in the same context.

VALLÉE: But if it's *really* important for a country, for a large power on earth, to keep something secret like this. . . .

HYNEK: If they can, yes. All right, you're knocking down the argument. I would say it is difficult to keep any secret very long.

HASTINGS: Has there been any example of that, of a secret project of that magnitude?

HYNEK: Well, of course, we don't know. And we wouldn't know it.

HASTINGS: The Manhattan Project.

HYNEK: The Manhattan Project, that's a good one. It could not have been kept quiet much longer, because a number of us were being quite conscious that physicist So-and-So had gone off on a mission . . . where was Commander Parsons? I asked one day and I got shut up right away. Well, he was one of the guys on the Enola Gay. He was the guy I was dealing with every week, and suddenly Commander Parsons disappears. "Well, he's on a special mission." And too many people were being segregated at Oak Ridge and places like that. You couldn't keep that too long. Sooner or later someone's going to find out why he is there and why these people are disappearing, so the Manhattan Project could not have been kept secret much longer. . . .

VALLÉE: That's because there were lots of people, but suppose you could have done it with fewer people.

HYNEK: Oh, yes, if you could do it with ten people, then you could keep the secret for a hundred years. But in the UFO thing, with the numbers of sightings, it implies a large operation, I think.

HASTINGS: Can you estimate the size or the magnitude? Compare it to something we know? You mention this in your lectures. Did you say it compares to ten Apollo launchings a day or something?

VALLÉE: That's rather important. That's an offhand thing, but somewhere we ought to get a kind of order of magnitude.

HYNEK: All right. I think it's important in talking about the magnitude and scope of the problem. It happened once more at my last lecture in North Dakota. I asked, after my lecture, "How many of you have seen UFOs or have a close associate who has?" Eighteen percent of the audience raised their hands in that particular case. And then when I asked, "How many of you reported it?" I got one hand. Now this has happened so damn many times that there's just no question in my mind that there is a reservoir of unreported cases, most of which—four-fifths of which, at least—are IFOs (identified flying objects) not UFOs, but that's all right. It still shows that, the United States being a relatively small part of the world, there must be, I would say, probably *a hundred sightings a night someplace around the world.*

HASTINGS: Caused by a separate incident, each one? Or is it six or seven people seeing the same thing? I mean, has that been worked out?

HYNEK: That's a very good point. I would say a hundred separate events. UFO sightings tend to be very local things. It's not like sighting an airplane, you know. When an airliner leaves Kennedy going to San Francisco, it is seen here and here and here, because it has horizontal trajectory. We have very few cases in which the thing is seen in Keokuk first and later in Davenport. The thing, whatever it is, seems to have a much more vertical trajectory.

HASTINGS: The question occurred in my mind that if you have something that moves as incredibly fast as these things appear to move, they can be seen by a lot of people in a very short time, and one conclusion might be—when you can plot all of them out—that there are a lot of these things up there. Either that, or everyone's seeing the same thing in a very short time, in a wide geographic area. The evidence seems to be that it's not a straight-line trajectory in most cases; so you can't conclude that they're all seeing very few objects. Then what about Michel's study of the French sightings falling on straight lines? Was there any inference drawn that a UFO was seen at point A at time A and point B at time B, etc.? And would it have been the same object along the whole line?

VALLÉE: No, it would not have been the same object. There were only maybe half a dozen cases in France where the same object was seen first here and then there, and then somewhere else. But there is at least one well-proven case in this country, the Bismarck case, where an object was seen coming over all the way from South Dakota to North Dakota and was tracked.

HYNEK: This is one of the very best cases, and it has never been publicly documented. You see, Jacques, we don't have any cases in which, if there were four people or five people in the group, only two or three saw it and the others didn't. Now, in the psychic cases, you hear that only one person saw an apparition and the other person didn't at all . . . sort of like one person seeing a ghost and the other person not seeing it. In the great, great, great majority of UFO cases, if one person sees it, they all see it. It attests to a solid reality. In a room of five people, if three people said, "We saw it," and two people said, "We looked and couldn't see it," that would be damning. But that doesn't happen. I think that when we describe the UFO phenomenon, this is one of its characteristics.

HASTINGS: Do those cases get reported very frequently, compared to the others?

HYNEK: You mean, if they hadn't all seen it?

HASTINGS: Yes. If three people see it and two don't, I would think the three people would hesitate to report it. They would say, "We had a weird experience. I saw something, but So-and So didn't."

HYNEK: That's a big selection factor. They probably don't report it.

HASTINGS: We're talking about . . . what are the other arguments you use against its being done by somebody in his backyard laboratory? One was the difficulty of secrecy.

HYNEK: All right. And the other is that, in general, a secret device is not widely tested over a hundred different countries! It's usually kept secret. You don't show it over populated areas. Of course, you can say that UFOs tend to land in isolated areas, but then there are so many isolated military bases where the thing could be kept! Why choose Delphos, Kansas? Why choose Marlin, Texas, or some of these places to test them, when you could set aside the Four Corners and test it there, and keep it a perfectly secret thing? And the third is a moral argument. Suppose we had the technology and had had it for twenty-five years with the right-angle turns, the rapid takeoffs—all the things that are reported—we would not be sending astronauts to the moon and using outmoded chemical fuels!

VALLÉE: It really doesn't have to be the U.S. government doing it.

HYNEK: Right, that's true, it could be from another country, but it would have to be some government.

HASTINGS: You would have to assume, also, implicitly, that somebody had discovered this kind of technology, technology which we have no way of explaining.

HYNEK: Yes, Joe Blow, well, maybe a mad genius; maybe there is a group of mad geniuses in Kansas City.

HASTINGS: Maybe Tesla lives![5]

HYNEK: Yes.

VALLÉE: Or an occult society.

HYNEK: Well, generally, people in occult societies don't even know that water is H_2O. I mean they disown the physical world and don't know how to work with physics. But, on the other hand, they may have superseded physics; they may have discovered other ways of doing things—that's what you're saying?

VALLÉE: Much of technology is very linear. The reason we have radio telescopes and the reason we have radar is because of World War II—because someone was already an expert with radio waves and happened to think of the right configuration along the line of technological development.[6] If somebody, say way back in the 15th century, had found an entirely different science on that basis....

Suppose a group had investigated along the lines of rational thinking and finally contacted other forms of consciousness? They would have kept it secret, because their leaders were worried of the consequences of releasing it in an unprepared world. Why didn't we keep secret the fact that $E = MC^2$? Why didn't we keep that secret? Because we were unwise. Suppose they were wise, and they did keep their findings a secret. And now they are making UFOs in their backyard. How is that for a wild theory?

HYNEK: Yes. Why didn't we keep $E = MC^2$ a secret? Then there would have been no A-bomb.

HASTINGS: That would explain that they're playing practical jokes, such as picking up rocks and putting them in a bag. . . .

---

[5] Nicola Tesla, a genius and inventor who designed the first generators of alternating current.

[6] See Appendix F.

VALLÉE: What's wrong with my scenario, then? Well, I'll tell you what's wrong with that; we know that occult organizations in the U.S. are usually groups of crackpots, idealists, and little old ladies. But then, maybe there are groups and occult societies that really have kept themselves secret. If so, a group like that could have the motivation and the means to manipulate public opinion on a grand scale.

HYNEK: If psychic projection is a fact, then these things could be kept secret and could be manifesting on a psychic plane. I don't know. We're just a couple of scientists trying to explore this for a second—but look at the impact that's happening. We've got the Uri Geller controversy; we've got books being written on extrasensory perception; books written on out-of-body experiences—all of these things are pointing or signaling that there's a reality that the physical scientists, the Condons, the Menzels, aren't at all conscious of, but exists!

Psychic Projection

HYNEK: Now, as part of our hypotheses when we're examining the panoply of hypotheses, we should take into account the paranormal phenomena that are being written about and apparently being experienced. Should those psychic claims be true, it opens up another can of worms. Then the problem essentially is solved; that explains why UFOs can make right angle turns, that explains why they can be dematerialized, why sometimes they are picked up on radar and sometimes not and why they are not detected by our infrared equipment. All that. But that's dangerous territory to tread.

VALLÉE: Why does it explain all that?

HYNEK: Well, because psychic projections aren't picked up on radars, that I know of.

VALLÉE: That's right, as far as we know they aren't, but UFOs are.

HYNEK: Some are, some not—there are many times when UFOs aren't picked up.

VALLÉE: There are ways of making an airplane invisible to radar.

HYNEK: Yes, well, I'm not sure that there are, that's one of the things that they're certainly trying very hard to do. But if they had that, then all planes, or all military planes, would be fixed up so they couldn't be picked up on radar.

HASTINGS: Invisibility is one of the occult abilities.

VALLÉE: I understand what you're saying, Allen, but it's not clear to me that it explains it. I mean when you say it explains everything, I don't agree with it.

HYNEK: It certainly would explain the apparent violation of the laws of physics, because psychic projection doesn't have to obey Newton's law of gravitation, it doesn't have to obey $F = ma$. To that extent it would get around it.

VALLÉE: It has to obey $F = ma$ if it has mass.

HYNEK: Psychic projection probably doesn't have mass.

VALLÉE: Then it doesn't explain the eight hundred cases of traces left on the ground by UFOs that Ted Phillips has collected. You said in the beginning that those are the things that any theory must explain, right? If it doesn't explain it, why would it be a good theory?

HYNEK: All right, why do poltergeists move something and yet don't appear on radar? Do they, the poltergeists, have that physical reality? And yet they have physical effects! In other words, we have a phenomenon here that undoubtedly has physical effects but also has the attributes of the psychic world.

HASTINGS: It's like a triangulation. From this field, we have certain phenomena; and from this viewpoint, we have others.

HYNEK: Well, we have a similar thing in the nature of light. Light behaves like a wave and light behaves like a particle. There are certain experiments in which light absolutely behaves like a bullet, and you couldn't possibly explain it by the wave model. In other experiments, it behaves completely like a wave, and the particle model won't work. UFOs have a similar duality.

VALLÉE: Okay, then let's consider the fourth hypothesis. Here I would see the psychic phenomena playing a role. Man is capable of psychic manifestations because he is a psychic being, just as he is a physiological being, and we could speculate that many latent psychic capabilities exist in us but have never been needed by man. Is that why they haven't been developed? Maybe we are only now becoming aware of UFOs and all those other things that we've sighted because they correspond to a new stage in our natural evolution, whether it was preprogrammed by nature or not. A good example showing that there are unusual human abilities is the case of the lightning calculators. These are people to whom you can give three twenty-five-digit numbers and ask them to add up the cubes of those numbers and they will write the answer instantly.

HYNEK: All right, that's good. I like that.

VALLÉE: That's an ability that can be demonstrated; they don't have to be in a special state, they don't have to take a special drug, it's repeatable. They can do it eight hours a day, and there are dozens of people like that in the world, some of them in insane asylums. Okay, if they can do it, why is it that we can't *all* do it?

I think the answer is that nature has never needed an animal that could do that and, therefore, it hasn't developed it, but it's a latent ability. It just wasn't needed for survival. Now, if it should happen that only those people who can compute the cubes of twenty-five-digit numbers in less than one second could survive, then perhaps the human race as a whole would acquire that capability, because all those that couldn't do it would be eliminated.

HYNEK: This is rather amusing, because my predecessor, the first director of the Dearborn Observatory, was a lightning calculator. He was a child prodigy.

VALLÉE: Did you tell Arthur that he was also the man who "explained" the 1897 airship?

HYNEK: Yes, he was the guy who ... in fact, you should go ahead and tell the story.

VALLÉE: In 1897, there was a mysterious airship seen by hundreds of people in Chicago, so reporters went to the director of the Dearborn Observatory. They asked him what it was. He said that he was having dinner, so would they please come back in half an hour, or words to that effect. The thing was still out there when the reporters came, but a half hour later it had gone away. So, he went outside, saw a big, bright thing in the sky and said, "Oh, that? It's the star Alpha Orionis!" Thereupon, he went into the dome, opened it, turned the telescope on the star, and triumphantly identified it as Alpha Orionis by its azimuth and elevation. By that time the airship had gone somewhere else.

HYNEK: Yet, in the instruction book for the HP45 computer, it starts off with the story about how he was given a fantastic mathematical problem and it said how he tore at his pants and he squirmed and he was in agony, but in less than a minute he gave the answer. Now, I think that's a good point because we have these people who can add up the numbers on the boxcars as they go by. We have these prodigies and they're just as unexplainable as the UFOs. Nobody can explain them.

As a matter of fact, I think I should tell the story of when I was assistant dean of the graduate school of Ohio State, and they asked me to sit in on the doctorate examinations of other schools. I was asked once to sit in on a PhD examination in psychology and this candidate was so cocky—psychologists sometimes are, you know—so cocksure that when it came my turn to ask a question, usually the dean says, "Well, I'll pass," but I didn't want to this time. So I said, "Well, this is completely out of my field, but I find this extremely fascinating and the candidate seems so knowledgeable ... therefore, I would like to know what happens in my mind when I imagine that I am having tea in a house in Isfahan while looking at some beautiful flowers in a vase, and listening to a symphony orchestra." The candidate drew a blank on that and the adviser had to admit they had no idea of how it worked.

Interlocking Universes

VALLÉE: What other wild hypotheses could we make?

HYNEK: There could be other universes with different quantum rules, or vibration rates if you want. Or our space-time continuum could be a cross-section through a universe with many more dimensions.

HASTINGS: This may not be a sensible question, but I don't understand; if they're going at different rates, how could anything ever cross over?

HYNEK: Well, the probability might be very slight, but in the same sense there is no transition permissible in ordinary matter between quantum number zero and two for instance. The transitions normally take place by increment of one. But the zero to two is permissible. It has a *very low* probability of ever happening, but it *can* happen. We say it's "forbidden" only because it has such a small probability of happening. If you have these interlocking universes, the chance of going from one to another is extremely small. But also there may be a trick for doing it. A "mind over matter" trick. Out-of-body experiences might be an example. If you can imagine what people describe as "astral projection," it might be an analogy. Ordinarily, the transition is forbidden.

HASTINGS: Maybe I'm thinking of it in the wrong way. I'm thinking of two groups of people dancing, one at a certain rate and one at another rate. Then the slower dancers can't join the fast ones unless they dance fast too. So how can you cross over?

HYNEK: You're saying that they won't meet because these are going slow, and they have to catch up. But that's a spatial argument. Think what a hard time you would have proving to an aborigine that right now, through this room, TV pictures are passing! Yet, they're here. You have to have a transducer to see them—namely a TV set. Well, in the same sense *there may be interlocking universes right here!* We have this idea of space, we always think of another universe being *someplace else*. It may not. Maybe it's right here.

VALLÉE: But the out-of-the-body experience involves going "there" and acquiring knowledge, as in remote-viewing experiments.

HYNEK: Yes, remote viewing of other physical locations.

VALLÉE: But here in the case of UFOs, we're dealing with something that, for a while at least, is *physically* in our universe. We can touch it, we can smell it, we can feel heat from it, we can even see the traces.

HYNEK: Yes, in the same way that a poltergeist can produce very real physical effects.

VALLÉE: So that's not the usual out-of-the-body experiences as they are described in the classical literature. They do not result in material inter-penetration of the universe.

HYNEK: But you have had experiences where a person wakes up and sees a person in his room and they are almost tangible, or they say that they felt sort of a breeze or they felt a light touch on their skin or something of that sort. Remember the presence in the room at the time of the UFO encounter in the Ely case.

VALLÉE: So it's physiologically real, but there is no evidence that it's physically real in that case.

HYNEK: That's right, that's correct. There is a physiological effect. But reality, when you come right down to it, is a physiological construct.

Conclusion

We stand at the edge of reality. More than ever we must consider with an open mind the phenomena that puzzle us. But an open mind is not an uncritical mind, and the main thread through the nine chapters of this book has been our trust in future science, our belief that it will develop an accurate understanding of the facts we are confronting. For this to be possible, we have shown that some attitudes and many biases would have to change. The study of UFOs is an opportunity to move toward a new reality, a means of increasing the borders of our awareness.

The edge of reality is also the edge of knowledge. But beyond this edge is another science and another knowledge. If this book has provided a few glimpses of it, we have accomplished our purpose.

APPENDIX A

A Sighting in Saskatchewan

On the misty Sunday morning of September 1, 1974, about 10 A.M., thirty-six-year-old farmer Edwin Fuhr began harvesting his rape crop in a field located some fifteen hundred feet south of his home.

The farm, located five and a half miles northeast of Langenburg, Saskatchewan, is eight miles west of the Manitoba border and 130 miles north of the North Dakota-Canadian border and 136 miles east-northeast of Regina. This region of Saskatchewan is in the plains region, devoid of elevations of striking prominences. It is open, flat country, primarily a wheat-producing region, and the soil is a rich loam from ten to twenty inches in depth, resting on a clay subsoil that retains moisture.

At the time of the sighting, the weather conditions were: overcast, low cloud ceiling, light showers in the area, temperature forty-three degrees, and the wind northwest at twelve miles per hour. Ted Phillips, one of the most experienced investigative associates of the Center for UFO Studies, conducted an investigation on September 21 and 22, although local investigations had already been made and the Royal Canadian Mounted Police (RCMP) had taken several photographs of the affected area.

After Fuhr had been swathing for nearly an hour, he was closing on a slough located at the south end of the field. Looking up to check his position and slowing his swather to a crawl, he saw a metal dome about fifty feet away, apparently sitting in the grassy area between the slough and the crop area (see Figure 17). He stopped the swather, descended, and walked to about fifteen feet from the object. As he approached the metal dome, he noticed that the grass around the base was moving and that the object itself was spinning at a high rate. Fuhr understandably became quite frightened and backed away—he did not take his eyes away from it—until he had backed all the way to the swather, which was still running at full throttle.

He moved behind the swather, climbed up to the seat, and viewed the spinning dome from a higher vantage point. Now he also saw *four more* metal domes, all the same size, and all spinning. They seemed to be hovering a foot or two above the ground, and he noted that the grass was in disturbed motion at the base of each object. Two of the objects were near to each other, with all objects, however,

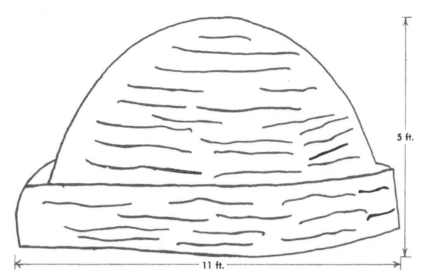

Figure 17 Sketch of UFO as drawn by Edwin Fuhr.

arranged in a rough semicircle around the slough. He could hear no sound, possibly because the swather was still running.

After what seemed to be several minutes, the objects suddenly ascended into the air in a step formation to a height of about two hundred feet. They stopped, and a puff of dark gray "vapor" came from the exhaust-like extensions located at the base of each object, accompanied immediately by a downward gust of wind which almost knocked Fuhr's hat off and flattened the rape that was standing in the immediate area. The objects were now positioned in a perfect line. After remaining stationary for perhaps one or two minutes, they suddenly ascended into the low cloud cover and disappeared.

Fuhr later learned that, at about the time of his sighting, cattle in a nearby field were bellowing and had broken through a fence in four places. After the objects had disappeared, Fuhr again approached the landing area and found five rings of depressed grass which had been swirled in a clockwise direction in all cases. The grass was not dead and had not been scorched or burned.

Now what does one do with a UFO report like that? One can dismiss it as the product of a deranged mind (especially since there was but one witness to the actual sighting), but there is no evidence whatever that Fuhr was not completely sane, both before and after the event. Furthermore, how does one explain the rings? One can always say "hoax," but one should have some evidence before making such an accusation, and there isn't a shred of evidence in that direction.

Let us proceed to the taped interview of Fuhr by Phillips (September 21, 1974) and allow the reader to evaluate from excerpts of the transcript the situation for himself. All we present here is an example—one of a very great many—of UFO reports which fit the time-honored description of "incredible tales told by credible persons." These are our "raw data." We don't like the nature of the data any more than any critical reader; we too would like movies of the event, and at least a dozen witnesses, at least one of whom was a PhD in physics and another an expert in optics, meteorology, and mechanics. Better, a chosen panel of experts brought to the scene with cameras and other scientific equipment, much as scientists assemble well in advance of a total solar eclipse. But the UFO event is unscheduled; it appears to prefer isolated areas, it lasts only a short while, and only rarely does it leave tangible evidence of its having been there. That is the situation we are faced with. Do we pay no attention because we don't have infrared photographs, physical samples of the craft, and electrocardiograms of the witnesses? If so, we are no scientists, because a scientist should be curious about unexplained phenomena; they can lead to breakthroughs in our understanding of the universe around us. But on to the interview. . . .

Taped Account of the Event, Recorded 9/21/74

FUHR: I had about an acre and a half left on this end. I was coming up very slowly at about a half mile per hour. I was about seventy-five feet from that slough, and I had to slow down because the rape was laying flat. I was about fifty feet from that object, see; I knew I had to turn around and I looked up and I saw that, I call it a goose blind; I thought, "What the hell is that gut doing in that damn slough?" I got off and walked up to it, not thinking a thing about it. I had jumped off the swather on the high side [N], no steps on it, walked around the swather, moved toward it. I walked to about fifteen feet of the circle, to about here, and the rape was standing up here yet [four or five feet high], I stopped, this wasn't swathed yet. I was just standing looking at it. I couldn't figure why the grass was wiggling around. The grass was standing up here and it was moving; I couldn't figure what the devil it could be.

I stood there about two minutes and thought, "Look, the whole damn thing is turning." I must have stood there at least two minutes, and I couldn't figure out what the devil it could be, so I backed up slowly to the swather. I never turned my back on it once, I just backed up slowly to the swather, got around behind it, and got up on the left side of it. When I got on, I sat down in the swather and then I saw those four on the lefthand side of me. They were all revolving, all four of them. I sat there like I was froze, I couldn't move nothing. I didn't know what the devil to do. I sat there for, it could have been fifteen minutes or twenty minutes, I don't know; it could have been even less than fifteen minutes.

They all went up, straight up, to, I would say, about two hundred feet, and they stopped at that distance. If you had winked, you would have missed the takeoff from the ground to the point where they stopped [two hundred feet]. When they hit the two-hundred-foot mark, they stopped spinning and a vapor floated out, an exhaust. The exhaust was only about six feet long, like a vapor; it was from two ports at the bottom of it.

PHILLIPS: Now, you could see these ports, could you see holes or. . . .

FUHR: You could see they were about twelve-inch-diameter ports, and they were all like that. They were in a formation like a step, the lower was the last to go up. And after that, I would say only a second, there was a downward wind, a pressure that flattened the rape that was standing, and I thought, "Oh, hell, here goes my crop," and there was just a downward wind, no twirling wind; I had to hold onto my hat. After about two minutes, they were gone. It took just seconds to get to that height [two hundred feet] and then they were just standing and after that, into the clouds and they were gone. I sat two minutes in the swather to be sure they were gone.

It was overcast and it was raining and I would say the temperature was about thirty-eight degrees. I got off and went to the swather to see if it was warm; it was cool. If the machine was hot, it should have been steaming with the cool rain falling, but it wasn't. The color of the machine was like a brushed stainless steel. It was rough and you could see while revolving that it was sort of grooved all around, you could see kind of grooves, they were darker gray. A dome shape and the bottom part had that lip on it and it was a dark gray in color like it had been hot at one time, like steel that gets hot and cools off. To my knowledge, the whole thing was spinning.

PHILLIPS: Did it appear that the objects were about the same size as the swirled areas here?

FUHR: They seemed to be, I can't be sure; I was, I guess, kind of in shock. Now that the whole thing's over, sometimes I sit down and just wonder. "They could have been larger," I just don't know; after a while it makes you wonder, "Well, jeez, you must have seen them." People tell you all these stories and pump you so full of BS that you don't know if you are coming or going. So, I don't know; I'm damn sure they were all the same size, I couldn't tell any difference.

PHILLIPS: The one nearest you took off first?

FUHR: Yes, this one here took off first and then the one there and those two that were close together took off together and then the last one, and that's the way they were in the air too, in a step formation. When they got to that height [two hundred feet] and they were straight back, they looked like they were standing, and I thought, "My God, are they coming back again?" That's when I didn't know what to do. After that vapor, the wind and then into the clouds. The vapor came out as they stopped, you could see it for just a second, then the wind, the downward pressure. The vapor was a dark gray and toward the end [lower end] it was lighter and lighter, just disappearing.

PHILLIPS: How large did the objects appear to be?

FUHR: The top was about five feet and from straight across it was maybe eleven feet with that lip on. When I was sitting on the swather, I couldn't see the bottom part, I just noticed that all five of them were sitting there, and I was watching them at an angle, and I could see all of them at the same time. That one looked like there was something out of it, it looked like something was probing around in the grass. It was like, oh, I would say, the size of a fifty-cent piece, a probe it looked like and the grass was all twisted and you could see marks like something had jumped here and there, all over.

PHILLIPS: While you were looking at the near object on the ground, what did you think it was?

FUHR: I thought someone was playing a trick on me; I took it for granted. I have a neighbor who will play tricks, and when I saw it I thought, "What the devil's he doing now?" I thought it was one of those new metal goose blinds, and I thought, "What the devil could it be doing there in that slough, there's no geese or water in there?" And I thought, "Well, I'll walk up there and scare him," and it scared me instead. It didn't look like a goose blind then, it was revolving. If it had been on the ground or stationary, I probably would have walked right up to it.

PHILLIPS: Could you tell if it was on the ground?

FUHR: I couldn't see the bottom, but it was probably twelve or eighteen inches above the ground; it wasn't on the ground at all. The grass around it was always moving, it was moving steady.

PHILLIPS: When you were within fifteen feet of it, did it start to climb?

FUHR: No, it was sitting in that same spot, 'cause I backed up to the swather and I seen those over there, and they were all sitting at about the same height off the ground, about twelve or eighteen inches or so. They all seemed to be revolving at the same speed too, according to the way the grass was turning. They were revolving clockwise. When I backed up, I went slow back, I didn't turn my back on it, no way. When I sat down on the swather, that's when I saw the other four, and that's when I couldn't move. I know the swather throttle was wide open; I had never slowed it down. When I got close, my head wanted to go fast back but my feet didn't want to move.

When I got to the swather, I didn't know what to do, but when I got to the house, that was worse yet; I didn't know what to do. I wasn't going to tell them; they asked me, "What's the matter with you?" I told them I had seen something out in the slough; they wanted to know, "What did you see?" So I told them. They said, "No, no, you couldn't, you gotta be nuts." And I said, "No, come out and look, the swath is all down." When I told dad, he never said too much, he never said yes or no, but mom said, "No, that's impossible." I told dad to come and look; he wouldn't go by himself so I went with him. He looked, jeez, he couldn't believe his eyes, he went to each one, looked at all of them. Then we found that one spot on the grass where something had been out and you could see what looked like probes, where something had probed around. The grass was all tangled up. He was crawling around on his hands and knees all Sunday afternoon in that one spot there.

PHILLIPS: From the very first observation until they disappeared into the clouds, how long did you see them?

FUHR: The most it could have been was fifteen to twenty minutes; it could have lasted more or less, I can't be sure.

PHILLIPS: You have dogs, did they bark during the observation?

FUHR: Well, he barked, when was it? . . . *Saturday night,* they barked, the neighbors dogs barked too [the night before the reported event], they all barked at the same time Saturday night, about midnight. Then they barked about *three in the morning.* Jack, our neighbor, had a babysitter who was frightened because the dogs were barking, and when Jack came home, he said the dogs were still barking. On Monday night about 10:30, the dogs were barking. My dog had been out in the field area and he backed up to the house. The television was acting up about then too. *The dog wouldn't go into the field;* he usually follows me, but he wouldn't go out there. But the dogs were barking on Monday night and on Tuesday morning I found that mark there. When I heard the dog barking Monday night, I thought, "It couldn't be out there again; even if it is, no way am I going out there."

PHILLIPS: I understand that your neighbors cattle were disturbed. . . .

FUHR: Yes, the cattle were making a lot of noise that Sunday morning, the fence was broken in four places.

PHILLIPS: When they reached the two-hundred-foot level, did they line up right away?

FUHR: Yes, they all seemed to be just like man controlled; well, the way they took off they looked as if they were man controlled, to take off in a formation like that.... They were in a step formation, the lowest one was at the far end and when they were up there they looked like they were straight across when they were at that level. I don't think they were radio controlled, they landed just so many feet from the rape crop all the way around the field, each one. Now those two over there were real close together; they looked like they were only six inches apart.

PHILLIPS: When did you find [the other] site?

FUHR: The Saturday night after [September 14]; the dogs were barking again, and we found that one.

PHILLIPS: When you went to the sites after the objects had ascended, how did the grass appear?

FUHR: Well, I checked for burns but I couldn't find any. *The grass wasn't broken off, it was flat, pressed down.* It didn't seem different from the other grass except it was flattened; it wasn't dead or burnt or anything. Some sprouts are coming up there now, it's not dead.

PHILLIPS: After the objects left, you waited two minutes; then what did you do?

FUHR: After I looked at the marks, I continued to swath for quite a while because I didn't know how to go home and tell those guys at home; that was my problem. When I got home, they asked me, "What's the matter?" I was all pale in the face; I didn't say nothing. I washed and tried to eat, and I was trying to think how I could tell them about this thing. I went in for lunch at about 12:30.

PHILLIPS: When the objects ascended, did you feel any kind of sensation?

FUHR: None, I had to put my head back to watch them; no sensation, just that gust of wind, that downward gust of wind. I couldn't hear any sound because of the swather motor.

PHILLIPS: What do your friends think about your sighting?

FUHR: Well, I've had lots of calls from the news people, most of them seem to be really interested. Most of the younger people in Langenburg believe it, some of the older ones don't, but they don't believe the United States has landed on the moon. People say why don't you carry a camera? How the hell could you carry a camera on a swather?

PHILLIPS: What did the RCMP people think of the event?

FUHR: Constable Morier was really interested but the Corporal said not to repeat the story, he said, "Keep it under your hat for a while and don't report nothing."

PHILLIPS: Before you saw these things, did you believe in flying saucers?

FUHR: No, I thought it was a bunch of bull; I had never seen one, so why should I believe in them? God knows I do now.

Taped Interview with Edwin Fuhr's Mother

MRS. FUHR: I was just here from the church when Edwin came in ... he acted altogether different. He was sorta, you know, sorta worked up; he was so worked up that he couldn't hardly eat dinner. We only had lunch for dinner, so I said, "Why don't you eat the rest?" He said, "I don't want nothing, I'm not really hungry." I said, "Why?" And he said, "I saw something this morning." And his dad was sitting there and said, "What do you mean you saw something?" And Edwin said, "Oh, I can't even describe it to you." And I had a little bowl sitting on the table, a little stainless steel bowl, and he described it on that bowl. He said, "That's what it looked like, that stainless steel bowl."

PHILLIPS: So you were gone to church?

MRS. FUHR: Yes, we were gone to church and when we come home, his dad was home listening to the news on the radio. So I and Edwin's wife went to church that morning. Edwin was just getting ready to go out when we went to church, that was a little after ten. I didn't want to believe it, but he said, "Are you silly? I'm not going to tell you any stories; it is true." His dad went out with him to look at the marks, and Edwin came back pale as a ghost and said, "What next is going to happen?" I wouldn't go out there for five days, it was on a Friday, oh, they were fresh. I even dreamed about them.

Taped Interview with Constable Ron Morier, RCMP: 9/21/74

PHILLIPS: How did you first learn of the Langenburg report?

MORIER: Well, first of all, the fellow's [Fuhr's] brother-in-law lives next door here, and he is a good friend of the Corporal. I was on duty Sunday night and he phoned; he stalled a bit and I could see that something was on his mind. He asked me, first of all, if we had any reports of any mysterious sightings or anything like that. I said that we hadn't. He stalled then and wasn't going to tell me. Finally, he did tell me that Mr. Fuhr had seen something; at that time he hadn't seen the rings as it was dark when he heard about it. He was at the farm visiting, and he was told by his brother-in-law what had gone on ... that he had seen the saucers ... and he was skeptical as hell, but he thought that he would check with us. So I said, "No, but it sounds awful interesting; if he says there are circles out there, let's go out and have a look at them in the morning." I was off duty at the time, I figured it would be interesting.

I got up about eight o'clock and went out there and met Edwin, and I could see that he was still, it appeared to me, to be quite shaken, you know, about this whole thing. He was jumpy, and you could see by just looking at the guy that he had been scared. But he took us out there, and sure as hell there they were, five circles. I was skeptical too, but I was curious. There they were, I had never seen anything like them before and listening to his story and everything ... so I got on the phone and called the Corporal and told him what I had seen and told him that I thought that someone should get some pictures. So, they did; they came out shortly and they took pictures. *These pictures were taken after only about four people had actually seen the rings*: Edwin, his father, myself, and ____. So that's how I came about the report.

PHILLIPS: When you were there, they did appear to be quite fresh?

MORIER: Yes, they were. The thing that really stood out was that the grass was all flattened out in a clockwise fashion. I got down on my knees and put my face near the grass, trying to smell some kind of exhaust or afterburn or something like that. There was no odor to it at all. But it was really swished tight, you know; *I mean it was really flattened and matted together*. I noticed, going out again, as the days went by, that the sun had dried the grass and it wasn't as prominent as before.

PHILLIPS: So the pictures were taken twenty-two hours after ... that would have been September second?

MORIER: Yes, the second of September.

PHILLIPS: You know, in talking with people around town, and the CBC[1] did street interviews with local people, you get the impression that the local people feel that Fuhr is reliable and sincere. In your opinion, based on your discussions with him, do you believe him to be an honest man?

MORIER: Yes, I do.

PHILLIPS: Do you believe he is sincere about the event?

MORIER: Yes, I believe that he saw something, and I don't see why he would exaggerate what he saw. There is no way that this is a hoax. Just talking to people who know him, of course I wondered as to his credibility too, but I have spoken to different people too, including relatives of his, in-laws, etc., and they all believe that he did see something out there. They've all seen the circles, they are convinced that something was there, and I am too. *I think that there is no way that anything was wheeled in and out of that field* because there had to be some trace and you saw yourself in those slides that when the swather was wheeled in there and the pictures were taken that you could see the tracks clearly.

PHILLIPS: And there was no evidence of that in any of the areas?

---

[1] Canadian Broadcast Corporation

MORIER: No, sir, it was in the slough grounds; the grass was green and it was long and it was undisturbed except for the circles, and whatever was in there, it came out of the air and departed the same way, as far as I could tell.

PHILLIPS: So, to your knowledge, Fuhr did not report the sighting to the news media himself?

MORIER: No sir, it got out through other people.

A brief timetable of events follows:

August 31, 1974–September 1, 1974

Midnight–3:00	Neighbors' dogs barking, babysitter frightened.
10:00–10:15	Edwin Fuhr began swathing (according to his mother—"So I and Edwin's wife went to church that morning. Edwin was just getting ready to go out, that was a little after ten").
10:45–11:15	Observation of five discs on ground, ascending, disappearing into clouds. NOTE: cattle on neighboring farm bellowing, broke fence in four places.
11:15	Fuhr checked watch—time 11:15—he had remained on the swather "a good two minutes" after the objects had disappeared into the clouds.
11:15–12:30	Fuhr looked at marks, continued swathing—"after I looked at the marks I continued to swath for quite a while because I didn't know how to go home and tell those guys at home; that was my problem"—"I went in for lunch at about 12:30."
after 2:00	RCMP Constable Ron Morier received call from Fuhr's brother-in-law.

September 2, 1974

8:00 A.M.	Morier visits Fuhr farm, photographs taken that morning by RCMP.
10:30 P.M.	Dogs barking at Fuhr farm ... dog had been near field, backed up to house. Television was acting up at that time also.

September 3, 1974

A.M.	Ring #6 found.

September 14, 1974

P.M.	Dogs barking again.

September 15, 1974

A.M.	Ring #7 found.

APPENDIX B

A Sighting in New Jersey

June 6, 1974

Dr. Allen Hynek
Department of Astronomy
Northwestern University
Evanston, Illinois

Dear Dr. Hynek:

As we discussed on the phone, I am sending a copy of a report, written mostly for my own record, of my UFO sighting on the night of June 4th, 1974.

At first, I could not resist some humorous asides in the report because I did feel silly when I sat down at the typewriter. Even now, having accepted what I saw, I feel—I really can't describe how I feel.

Upon re-reading my own story, I don't think I would change a word of it.

Sincerely,
Robert J. LeDonne
ABC News
Editor, Special Events Unit

June 4, 1974

6:00 P.M. Routine day ends at office. Off to subway.

6:20 P.M. Depart Port Authority terminal on commuter bus to Woodcliff Lake, N.J.

7:10 P.M. Town councilman offers me lift home. We discuss local taxes. What else?

7:25 P.M. Wife has prepared steamed clams dinner.

7:45 P.M. Still plenty of light outside. Work in garden, planting crop of extra-early, extra-sweet corn, which raccoons ate last year.

8:45 P.M. Kids spot raccoon in woods, casing new crop. Getting dark. Mosquitoes biting. Wash up.

8:55 P.M. Drive half-mile for cigarettes, six-pack of beer, and head home for Mets game on TV.

9:05 P.M. One block from home, I sight UFO.

Okay, maybe it wasn't a flying saucer. No intelligent, middle-aged man with a generation in the news business would admit he saw one without the greatest reluctance and apologies. UFOs are old hat. Even the Air Force has closed Project Blue Book, its endless study of UFO sightings. Nevertheless, I have filed one telephone report and one written report with the local police. Now, two hours later, I want once more to set everything on paper, exactly as I saw it, before my memory starts surrendering to my innate skepticism. Perhaps this way I can make sense of it. Besides, it may turn out to be an interesting experiment in self-analysis. There are conflicting forces within my mind, each struggling to win out. My eyes have transmitted to my brain a vivid picture, which my brain keeps trying to reject or rationalize.

It was a dark night. Haze obscured the stars. The full moon was barely above the eastern horizon, and hidden behind a wall of trees. Suddenly, in the southeast sky, I spotted a brilliant oval of lights. I immediately thought of the Goodyear blimp in the distance, enveloped in bright yellow lights with one red light in the upper area. There was a touch of purple in the red.

(Even now, as I put this on paper, I feel vaguely uncomfortable. I have an impulse to hold off until morning. Sleep on it. Maybe I'll change my mind. No. I want a fresh, complete record now.)

Let's see. The Goodyear blimp. An advertising stunt? No, that doesn't compute. I slowed the car, stopped, stared. A plane circling? Too many lights. Not that. What then?

I put on the emergency brake, stepped out of the car, and watched the object approach. I was filled with curiosity at first. Then puzzlement. Then mild anxiety of a sort—I urgently wanted another car to come down the road so I could hail it down and commandeer a witness. None came.

When I first spotted the object, it apparently was coming toward me. Now it was doing a fly-by at tree-top level, perhaps a quarter-mile away.

I now was viewing it from the side. The reddish light was in the rear. The bright yellow lights were in a line. Very bright, almost white. The vehicle was dipped in a slight angle forward. Most baffling, the lights were revolving, from rear to front.

The trees along the road varied in height. Most of the older ones had been cut down long ago for a housing development. The UFO was sharply visible above the lower tree tops. But when it passed the older, taller oaks, it was obscured by the upper branches, obscured but not blacked out. Then, after seconds that seemed eternities, the UFO would emerge again, skimming over the black curtain of younger trees.

How big was it? That's impossible to say. If it was a block away, then it was the size of a helicopter. If it was a mile away, it's anybody's guess.

I seized on the helicopter idea for several reasons. The reddish taillight was where it should be. The vehicle was tilted slightly forward as one would expect with a helicopter. Sometimes, it would dip abruptly, then return to its normal attitude. Those weird ultra-bright lights revolving around the craft were really baffling. I considered a helicopter pilot pulling a joke, or possibly an optical illusion involving a copter.

Cross out the chopper idea. No sounds of engines. A chopper at that distance, at that height, would be deafening on a still night in a rustic suburb.

Okay, an optical illusion. A star? Several stars? Impossible.

Some other aircraft? No, nothing but a helicopter could move that slowly and stay aloft.

It was frustrating. The damned thing was as garish as a ferris wheel in the sky.

Several rooms in the house across the street were lit lip, and I was tempted to run over, bang on the door, and invite them out to watch. I swiftly discarded that notion. If the thing was gone, I would be in one hell of an embarrassing position at the least.

Finally, after watching all this for two or three minutes, if one can measure time in a situation like this, the UFO went behind still another tree top and was about to go out of sight. I got back in the car and sped home.

I didn't even consider calling any of the news media. UFOs were big news a decade ago. Today, nobody's interested. There have been far too many reports, fake photos, and little green men. Nevertheless, I realized that whatever it was—and I began toying with the idea of an elaborate hoax—someone else should have spotted it, too. The cops would be the first to know. Psychologically, I guess I needed reassurance, so I called the regional police radio. The officer said there had been no other reports of UFO sightings. I fully intended to drop the matter there without giving my name. But oddly, the cop on the phone did not treat me like a crank. Maybe because I didn't sound like one. I was cold sober and completely rational. The cop did not appear to consider UFO sightings out of the ordinary, and said the Air Force liked to check out the police reports.

I was amazed that such a ridiculous story should fall on receptive ears. This officer actually made me feel that it was a public duty on my part to file a report, and he offered to send up a local squad car. With some qualms, I agreed.

While the police were en route, I swiftly documented my story on a typewriter. When the cop appeared at my door, only ten minutes after my initial phone call,

I was tongue-tied and apologetic. I thrust the sheet of paper, said that's it, and then self-consciously tried to laugh off the incident. Somebody's putting us on, I said. It's got to be a hoax.

But this officer took me seriously. After reading my report, he said that another local patrol car had made a UFO sighting in our town recently. It was in the vicinity of a farm toward which my UFO could conceivably have been heading.

The irony is that when he told me that story, my eyes widened. *I was skeptical of the police sighting.* Maybe it was because I didn't expect anybody to believe my story, so how could anybody expect me to believe theirs? Frankly, I just don't understand myself in this situation.

In retrospect, I was reverting to my normal, possibly cynical, newsman's attitude. Maybe I have been conditioned over the years, programmed not to accept the unacceptable.

Still, I saw what I saw! And I was left to grapple with what I knew to be true and found impossible to rationalize.

After the patrolman had left, I described the incident in detail over and over again to my wife. At her insistence, we bundled the kids into the station wagon and drove back to the place where I had sighted the UFO.

With the exception of the UFO, everything was exactly as I described it. In fact, only in that spot could I have seen all that I did. Fifty feet back, my view would have been hidden by road-hugging trees. The same, 100 feet up the road. The trees that I said intermittently hid the UFO were right where I said they were. The lights were still on in the house across the street.

My wife must love me because she agreed I wasn't nuts. I wasn't so sure about the people across the street. The man of the house had opened the front door, wondering what in hell was going on. We returned home.

My daughter, age eleven, insisted I sign her autograph book, indicating that I had spotted the UFO that very night. (I wonder what she'll think about that when she gets older.) The two little boys thought it was great fun. Everyone bedded down for the night except me. I went to the den and my typewriter. Here, I am now.

Now, I begin to apply logic. (Logic, in this case, is what you use to make things make sense that don't make sense.)

Of all the aircraft that might have been mistaken for the UFO, only the helicopter was possible, because of the slow tree-top speed and other similarities already discussed. But how do you explain the ring of revolving lights? If they were attached to the tips of the rotor blades, they would be whirling so fast that it would look like a ring of light. If it were a helicopter pilot's hoax, why not pull it off over a more populated section where more people would be suckered in? And why no sound of the chopper engine?

Scratch the chopper again.

Alright, how about the failure of others to make sightings? Well, the UFO was flying very low over a wooded area. If I had not been in the precise spot where I sighted it, I wouldn't have seen it myself. Besides, there was no engine noise to attract attention.

As for the people living in the neighborhood, these are the outer suburbs with plenty of trees. At 9:05 P.M., the Graf Zeppelin could go down in flames, and most people wouldn't notice it unless it was televised live. That's the time when you're putting the kids to bed, mixing cocktails, watching television.

Let's say it was an exotic, secret Air Force aircraft. No way. A secret weapon so brightly lit would not be a secret for long, especially cruising around the metropolitan suburbs.

An "enemy" aircraft? Russian, Chinese, or whatever. Preposterous!

A hoax? I don't see how it's feasible or logical. People perpetrate hoaxes only when they want the hoaxes to be widely publicized and believed. It certainly isn't logical to "con" just one lousy observer: me.

So now in our pursuit of logic, we come to the alternatives we prefer not to believe, but delight in discussing.

Having virtually ruled out earth-spawned interlopers, we are left with (and I gag on this) the possibility of extraterrestrial observers, coming down for a look-see; less believable but more logical. Your imagination can run wild to fill in all the holes. We may not be able to produce a flying saucer, but "they" could. Perhaps not expecting to find a civilization here, these "visitors" would not originally have been concerned with the high visibility of their spacecraft.

Perhaps once finding that we did have a high civilization, they attempted to conceal themselves in low flights over what appeared to be relatively lightly populated areas. If a "visitor" did want to make closeup observations, obviously it would be in the form of guarded low-flight tours at night in rural areas, or heavily-wooded suburban areas. Certainly not the cities in day or night!

If *we* were visiting another inhabited planet, we certainly would not confine ourselves to the barrens of their polar caps, their deserts, their wastelands.

The point is that of all the logical possibilities I have considered, this last is the most logical: visitors from space.

Of course, I do not accept it. My mind is very rigid about this. I already have concluded, with all the scientific knowledge to which I have been exposed, that no other planet in our solar system is capable of producing any species of life comparable to man.

On the other hand, I am equally convinced that intelligent life does exist on other planets in *other* solar systems. The odds are astronomically in favor of this.

However, I am convinced, too, that time and distance, among other factors, forbid expeditions between star systems.

Thus, with logic, I can rule out flying saucers entirely. They simply are not logical; therefore they are not.

The fact that I definitely and irrefutably saw a UFO tonight makes no difference.

June 5, 1974

I became convinced today that my UFO had to be a helicopter, despite the fact that I heard no engine noise. I began calling people in the helicopter business. Strangely again, I got a good hearing. Most were interested, and tried to help figure out how a helicopter, under certain conditions, might look like the UFO I had described.

Yes, helicopters do have red taillights. However, the helicopter taillight is usually a flashing red beacon. My UFO's taillight did not flash; its taillight was reddish purple and not too bright. In fact, it was not as bright as red lights I have seen flashing on airliners passing high overhead.

Could there be lights on the helicopter rotor blade tips that might give the illusion of my revolving yellow UFO lights? The spokesman at one heliport said that in the past there had been experiments with lights on rotor tips, but that the tremendous centrifugal force of the rapidly turning blades shattered the glass. He vaguely recalled one company being successful and marketing such helicopter rotor lights, but he was uncertain about it, and had never seen one. Of course, lights on the rotor tips moving around so quickly would give the illusion of a solid band of light, which again would not resemble the slowly revolving lights I had seen on the UFO. Furthermore, if there were such helicopters on the market, they would be bound to stir up hundreds of reports of flying saucers, which just as swiftly would be explained away.

I gave up on the helicopter people. The more I talked with them, the more they were convincing me that the UFO could not have been a helicopter.

I called the Air Force. I was told its study of UFOs, entitled Project Blue Book, had long since ended. The Air Force spokesman said all UFO sightings should now be referred to the FAA, but warned that the FAA seemed to be referring them right back to the Air Force. I wasn't about to jump on that merry-go-round, so I didn't call the FAA.

The next time there's a full moon, I think I'll stay inside.

June 6, 1974

I suddenly remembered an old NASA acquaintance, Al Chop, who was involved in a book about UFOs. I traced Al to Downey, California, and talked with him on the phone. His conclusion was that my sighting was comparable with other sightings, although not necessarily all of them.

He recounted the story of the UFO and an Army helicopter on October 18, 1973. The UFO bore a reddish light that was on a collision course with the helicopter. As it drew closer, the pilot saw a sixty-foot-long, cigar-shaped grey object which hovered over them briefly. Suddenly, the copter controls went haywire and the craft was sucked upward rapidly from 1,500 feet to 3,800 feet. When the story was told on the ground, Army officials said there was no reason to doubt the truthfulness of the men inside.

So, my story wasn't so bizarre after all. At least, Al Chop didn't think so.

I'm beginning to be more confident. I am now willing to accept what I saw. I can't explain it, but I accept it.

I also believe now that the local police did see a UFO that resembled a flying "pan" two months ago. I am no longer a skeptic.

As I close this report, I must say I will never forget that spectacular sighting. It is now etched sharply in my memory. I have vigorously sought a rational explanation, and I have finally found one: I saw a UFO.

It's that simple.

And of all the possible explanations for it, I must grudgingly concede I would have to put this UFO in the flying saucer category, because nothing else fits.

APPENDIX C

UFO Sightings by Americans Have Doubled in Seven Years
Gallup Poll—Opinion Index Report No. 103, January 1974

An astonishing 11 percent of the adult population, or more than fifteen million Americans, have seen a UFO (unidentified flying object)—double the percentage reported in the previous survey on the subject in 1966. The figure then was 5 percent.

In addition, the latest survey shows half of those who are aware of UFOs (54 percent) believing that these flying objects—sometimes called "flying saucers"—are real and not just a figment of the imagination or cases of hallucination.

Almost everyone (94 percent) has at least heard or read something about UFOs. For something so highly publicized, this finding may, at first, not seem unusual. However, in terms of the history of the public's awareness of other incidents or events, this figure is extraordinarily high. In fact, this awareness score is one of the highest in the thirty-seven-year history of the Gallup Poll.

The same survey also shows nearly half of all persons interviewed (46 percent) believing that there is intelligent life on other planets. This represents a sharp increase in the percentage with this belief since the 1966 survey when the figure was 34 percent.

It is interesting to note that persons who believe in the existence of life on other planets are far more likely to believe that UFOs are real and not something imaginary. In fact, seven in ten of those who think there is life on other planets think UFOs are real.

Analysis of the survey data shows that UFO sightings are not confined to any particular population group. For example, college-educated persons are as likely to have seen a UFO as are persons with less formal education. However, a considerably higher proportion of sightings is reported in the Midwest and South than in the East and Far West.

In addition, persons living in small towns or in rural areas are more likely to report having seen a UFO than are persons living in the larger cities of the nation.

A Gallup survey in 1971 of top leaders in seventy-two nations found 53 percent expressing a belief in the existence of human life on other planets, while 47 percent ruled out the possibility. The survey was of leaders in science, medicine, education, politics, business, and other fields, selected by careful sampling methods from the *International Who's Who*.

APPENDIX D

Soviet Ufos

Reported sightings in the Soviet Union include the case of a pilot who wrote:

> In 1956, while accomplishing strategic ice reconnaissance on a TU-4 airplane (at 300–500 kilometers per hour) with the aircraft commander, Commander Nakhtinov, in the area of Cape Dzhuze [as transliterated] (Greenland) and emerging from the clouds into clear weather, we unexpectedly noticed that some sort of unknown aircraft whose shape resembled a large lens was moving parallel to our general course (180°). It was pearl-colored with undulating-pulsating edges. Knowing that American air bases are located in northern Greenland [word illegible] we first decided that this was an American aircraft of unknown design and, not wishing an encounter with it, we took off into the clouds. After flying for forty minutes along a course to Medvezhiy Island, the cloud cover ended unexpectedly. It was clear ahead, and we again noted this unknown aircraft along the left side. Deciding to examine the nature of this object in greater detail, we changed course sharply and moved to close with it. The unknown aircraft also changed course and flew parallel with us at a speed equal to ours. After fifteen to eighteen minutes of flight, the unknown aircraft changed course suddenly—it flew ahead of us and departed upward rapidly, disappearing into the blue of the sky. We did not discover any antennas, superstructures, wings, or illuminators on this disc. Neither exhaust jets of gas nor an inversion trail were noted and the speed of its departure was so great that this phenomenon appeared to be something supernatural.

Another report comes from a scientist who was flying as a passenger aboard a Tupolev 104A on flight C94 from Leningrad to Moscow. The date was July 12, 1966:

> At approximately 1725 hours, halfway along the route when the airplane flew over the city of Bologoye, a gigantic airplane burst out from beneath the bottom of the fuselage at a tremendous speed and took off at a right angle to the line of flight of our aircraft. It seemed to me that this was some type of tremendous bomber of a design unknown to me which moved away so that I saw it exactly in the "tail" although later I could not recall any details (for example,

the stabilizers) which are inherent to airplanes. At that moment I thought how poorly the airlines are organized, there could have been a collision!

Perhaps, I would have remained firmly convinced that I had seen a strange bomber if the flying object, approximately three kilometers from the airplane and somewhat behind it, had not changed its flight course suddenly (by approximately 90°) and began to move parallel to the course of our airplane, overtaking it and exceeding its speed.

When the object turned, it evidently decelerated and inclined so that I clearly saw that it was a disc and not a winged bomber at all. The curvature in the upper part of the object was clearly distinguished, emphasized by the shadow. When the flying object turned and leveled out its flight, it again assumed its former shape of a winged "bomber." It seemed strange that it was flying with its "wing forward." A black projection was noticed in its lower part. Flying past somewhat, the object plunged into a cloud "mountain." Its diameter was considerably greater than the length of the airplane in which I flew. Its color was silver-gray, similar to the color of aluminum, which is distinguishable against a background of white clouds.

APPENDIX E

The Technological-Cultural Gap

The following letter is typical of a certain class of communication received by the authors, and it is published here to demonstrate the extent to which ignorance—the technological-cultural gap—can carry an individual beyond "The Edge of Reality" (cf. Chapter 6).

21 March, 1974

To: Mr. J. Allen Hynek

Dear Mr. Hynek:

I read in our paper something about research into UFOs.

They are transports from Hell, With demons in them.

Here's what the Blessed Virgin told one mystic:

"There are many agents of Hell loosed upon the Earth. They are traveling in transports. Do not be won over to a false theory of life beyond the Heavens, other then [*sic*] the Kingdom of God! Know that it is Satan who sends these vehicles before you! They are to confuse and confound you!

"These objects that take flight across your earth are from Hell! They are only the false miracles of your times. Recognize them, my children, they are not a figment of Man's imagination, they are in your atmosphere and they will become more dominant as the Fight goes on for the souls. There have been divisions of armies set up to fight Lucifer, Lucial despicable Creature of darkness, who has set himself to destroy man. Man's souls, and take them to the abyss, the souls of my children."

She said that there is no life on other planets, only on Earth and the Kingdom of God; and Hell, thats [*sic*] in the center of the earth.

Sincerely, etc.
(Name Withheld)

APPENDIX F

Scientific Discoveries as a Linear System

A good example of the slow and linear development of scientific technology is the case of radar.

1887: Hertz proposes to use the reflection of radio-waves by objects for detection purposes.

1904: Hulsmeyer patents a collision-preventing device based on radio echoes.

1922: Marconi proposes to use radio echo to detect ships in fog.

1925: Breit and Tuve study the ionosphere with radio pulses.

1939: The U.S. Navy tests a radar set in battle maneuvers on the *USS New York*.

19 Dec. 1939: The Germans use the radar principle in battle against British planes for the first time.

In the words of recent authors (Heiss, Knorr, and Morgenstern, "Long-term Predictions of Political and Military Power," Mathematica, Inc., 1973), "There existed a period of *over 50 years* between the first formulation of an idea and its implementation, in an area of utmost military and technical importance. There are countless other such illustrations."

For further information or to report a sighting, you may wish to contact:

Center for UFO Studies
PO Box 31335
Chicago, Illinois 60631
cufos.org
infocenter@cufos.org

INDEX

A
abductions
  Hill, 89, 93, 94, 95
  Pascagoula, 83, 95, 96–97
Adamski, George, 65, 79, 127, 144, 186
  cultism of, 171–172
  Flying Saucers Have Landed, 145
  life and lies of, 145–148
Aerial Phenomena Research Organization. *See* APRO
Aldrin, Edwin, 64
Alvarez, Luis, 71–72, 156, 166
amateur UFO interest groups, 181–185
American Institute of Aeronautics and Astronautics (AIAA), 173, 184
Anatomy of a Phenomenon (Vallée), 4, 13n, 121–122, 179–180
angels, 126
animal reactions, 14, 25
  Saskatchewan, 216, 220, 221, 224
aphasics, 102
apparitions, 118–121, 127
APRO (Aerial Phenomena Research Organization)
  information on sightings, 73, 75–76
  on magnesium sample from Ubatuba, 67
  membership and publications, 181–183
  quality of data gathered, 162
Armstrong, Neil, 64, 65, 130
Asimov, Isaac, 6–7
astronaut sightings, 63–65, 130
Astronautics and Aeronautics, 173n, 184
astronomer sightings, 32n
astronomy
  distances between stars, 131–132
  mysteries in, 32
  predictions, 201
  public interest in, 136
  solar system, 35, 131
  Aurora, Texas, 148–149
  Australia, RAF case, 59–60

B
Ballester-Olmos, Juan-Vincente, 76, 123, 144
Bananeiras, Brazil, 19n
Battelle Institute, 155
Bean, Alan, 64
Beckman, Fred, 163
Bethrum, Truman, *Aboard a Flying Saucer*, 186
Bible, 126
Big Data, 77–78
biology, 203
Bismarck case, 207
Blue Book. *See* Project Blue Book
Blum, Ralph and Judy, *Beyond Earth*, 187
Bohr, Niels, 11
Boianai, New Guinea, 47–53, 56–57, 62, 153
books. *See* publications
Borewa, Annie Laurie, 52, 56
Borman, Frank, 64
Bowen, Charles, 4, 133
  The Humanoids, 190, 199n
brain, workings of, 101–103
Brazil
  Bananeiras, 19n
  Ubatuba, 67

C
Cade, Maxwell, 104–105, 118
Caetano, Paulo, 19n
Calgary, Canada, 78–79
Canada
  Calgary, 78–79
  Chesterfield Inlet, 8, 10
  Gleichen, 19n
  Saskatchewan, 215–224
cancer research, 77
Capella case, 140–143
Carpenter, Scott, 63
Carr, Otis T., 187
Castaneda, Carlos, 127
  Journey to Ixtlan, 120
Center for UFO Studies
  analysis of truck damage, 39n
  computerized data base, 73
  contact addresses, 239
  photographic analysis, 80
  publishings, 184
  purpose of, 18, 59, 78
  reports from credible people, 24
  Saskatchewan investigation, 215
  scientific credibility gap, 176
  scientists for studying UFOs, 69, 91
  UFO Central hotline, 93–94
Central Intelligence Agency (CIA), 158
Cernan, Gene, 63
Challenge to Science (Vallée), 4, 13n, 26n, 173, 180
Chesterfield Inlet, Canada, 8, 10
Chicago, Illinois, 1897 case, 148, 211
children, fantasies of creatures, 122–126
China, UFO reports in, 55–56
Chop, Al, 230–231
Clarke, Arthur C., 146n
  The City and the Stars, 201
Clearwater, Florida, experiment, 137–140
close encounters
  cases, 15, 24
  catalog of, in Spain, 76–77
  statistics on, 28–30
  See also Hill, Betty and Barney; Pascagoula, Mississippi
  Close Encounters of the Third Kind (Spielberg film), 2
Collins, Mike, 64
color changes of lights, 14
  Boianai, 49
  Ely, 40–41
  Long Island, 32n
  See also lights on/from UFOs

Index 241

communication with UFO occupants
  Hill, Barney and Betty, 87–88, 93
  Mr. Masse in France, 204
  possible future scenarios, 192–193
  thought projection, 69
  waving, 52–53
computer study, 73–78
  Condon Committee, 173–174
  UFOCAT, 74, 173
computers, development of, 176
Condon, E. U., 65n, 172–175, 181
  interested only in American sightings, 192
  lead not followed, 135
  as skeptic, 72
Condon Committee
  behind the scenes, 172–175, 187
  creation of, 163, 164
  examination of magnesium sample, 67
  failure in informing public, 34–35
  Hynek's research on UFO reports, 181
  in hypothetical scenario, 192
  [possible] governmental cover-up, 71–72
  research failure, 80
Condon Report
  1967 conclusions, 1–2
  change of climate over the years, 161–162
  Final Report written at beginning, 174
  lack of science, 72
  referenced in research, 23
  republish it backwards, 181
  Roach's closing remark, 65
  Scientific Study of Unidentified Flying Objects, 65n
Congressional hearings, 60, 158, 164, 185
Connersville, Indiana, 24
Conrad, Charles, 64
Conrad, Pete, 64
consistency of data. *See* UFO reports, consistency of
contactees
  changing viewpoints on, 164–165

credibility of, 164–165
  hoax promotions, 149
  stories of, 186
  See also Adamski, George; Mekis, Karl; witnesses of UFOs
Cooper, Gordon, 63
cover-up, by government, 70–73
craft (size and shape)
  Army helicopter encounter, 231
  astronaut sightings, 63–65
  basic characteristics, 13–14, 25
  Boianai, 49–52
  Chesterfield Inlet, 8, 10
  Connersville, 24
  Dzhuze, 235
  Eagle River, 128
  Ethiopia, 133–134
  Hill, 84, 87
  Leningrad, 235–236
  northern US Plains, 108–109, 111–112
  Pascagoula, 97
  photographic evidence, 79–80
  Saskatchewan, 215–216, 217–219
  Torino, 60
  Woodcliff, New Jersey, 226–227, 230
crashes of UFOs, hardware from, 66–67
Creighton, Gordon, 4
Cronkite, Walter, 2
Crowley, Aleister, 121
Crutwell, Norman, 47–52, 56–57
  "Flying Saucers Over Papua," 47n
cultism, 171–172, 195

D

Damon, Texas, 68
The Day the Earth Stood Still, 69, 129
Detroit Free Press, 193
Dexter, Michigan, 163–164, 172
Dzhuze, Greenland, 235

E

Eagle River, Wisconsin, 122, 127–130
"edge of reality," 7, 10n, 95, 118, 151, 213, 237

Edwards, Frank, 185
Einhorn, Ira, 166
Elementals, 121
Ely, Nevada case, 18–22, 31, 37–45, 194, 213
Emenegger, Robert, *UFOs, Past, Present, and Future*, 65n
Ethiopia case, 133–134
European cases, 166–168
  See also France; Spain
Everglades, Florida, 133
evidence of UFOs
  circles in Saskatchewan field, 215–217, 220, 221, 223
  confiscated by Men in Black, 67–68
  environmental, 24, 25, 53
  sketches as, 81
  See also photographic evidence; witnesses of UFOs
extraterrestrial hypothesis (ET), 14, 54–55, 195, 196–198

F

Fawcett, G., 63
fear of ridicule, 3
  Chesterfield Inlet, 10
  as deterrent for reporting, 12, 30
fear of witnesses
  Barney Hill, 94
  Chesterfield Inlet, 8
  Damon, 68
  Ely, 21, 42
  under hypnosis, 98–99
  Saskatchewan, 219–220, 222–223
Festinger, Riecken, and Shackler, *When Prophecy Fails*, 186, 187
Flammonde, Paris, *The Age of Flying Saucers*, 190
flight and movement of UFOs
  basic patterns, 15, 25
  Eagle River, 128
  Ely case, 20–21, 40–43
  Ethiopia, 133–134
  Hill, 87
  northern US Plains, 108, 111, 112
  Saskatchewan encounter, 215–219
  Torino, Italy, 60
  witnessed by astronauts and pilots, 63–65

flight and movement of UFOs *(continued)*
  Woodcliff, New Jersey, 226–227
  See also trajectories
Florida
  Clearwater experiment, 137–140
  Everglades, 133
Flying Saucer Reader, 190
Flying Saucer Review (FSR), 4, 19n, 47n, 133n, 184, 187
Fontes, Olavo, 94
Ford, Gerald, 164
Fort, Charles, 188, 189
  The Book of the Damned, 188
Fowler, Ray, *UFOs–Interplanetary Visitors,* 183
France
  investigations, 4, 55–56, 102–103
  investigations by Gendarmes, 58–59, 60
  Lumieres dans la Nuit, 183
  magnetic detection, 81–82
  Minister of Defense on UFO phenomenon, 59, 92, 176
  patterns of data, 194
  published reports, 160, 167
  scientific recognition as end of research, 185
  statistics on UFOs, 29
  tradition of "Lutins," 201
Franklin, Benjamin, 34, 36, 201, 202
Fuhr, Edwin, 215–224
Fuller, John, 89
  Incident at Exeter, 187
  Interrupted Journey, 90, 187
future scenarios of UFO developments, 191–195

G

Galley, Robert, 59, 92, 176
Gallup Poll, 31, 175, 233–234
Garrett, Larry, 83–85, 90–91
Geller, Uri, 61, 98, 209
geological fossils, 36
Gill, William Booth, 48–53, 56–57, 153
Gleichen, Canada, 19n
Glenn, John, 63
Gordon, Richard, 64
governmental cover-up, 70–73
Greenland, Cape Dzhuze, 235

Gross, Loren, "The UFO Wave of 1896," 148n
Guerin, Pierre, 4
Gugliotto, Lee, 32n

H

Hanlon, Donald, 149
  Airships Over Texas, 149n
Harder, Jim, 171
Hardin, Captain, 155–156, 159, 169
Harris, Robert, "The Real Enemy," 132
Hastings, Arthur C., viii, 1, 6
  research on UFOs, 22–37
  See also interviews with Arthur Hastings
Hawkins, Gerald, *Stonehenge Decoded,* 189
Heilbruner, Robert, 135
Hickson, Charlie, 86, 96, 97, 99
  See also Pascagoula, Mississippi
Hill, Betty and Barney
  close encounter, 83–92, 95, 99
  comments on Pascagoula case, 93–94
  loss of time, 84, 88, 118
  mentioned, 33, 186
Hillenkotter, Admiral, 158
hoaxes, 7, 23
Air Force experiment over Clearwater, Florida, 137–140
Air Force labeling of reports, 160
Capella case, 140–143
George Adamski, 145–148
Martian buried in Aurora, Texas, 148–149
Venus invasion, 150–151
Hopkinsville incident, 94
hostility from UFOs, 133–136, 191
  Ethiopia, 133–134
  Everglades, 133
  hostility from witnesses toward UFOs, 94, 193
  northern US Plains, 110, 113–114, 115, 134
  Walesville jet fighters, 134–135
Humphrey, Hubert, 175
Hynek, J. Allen
  as consultant on Project Blue Book, 155–171

letter from New Jersey sighting, 225
letter from religious skeptic, 237
photograph of UFO, 57
research on UFO reports, vii–viii, 1–2, 4–6, 27, 143
on satellite tracking program, 65, 158
as skeptic, 4, 5, 72
The UFO Experience: A Scientific Inquiry, 6, 13n, 180–181
See also interviews with Allen Hynek
Hynek, Paul, vii–viii
hypnosis of witnesses, 83–106, 118
  close encounter investigations, 83–94
  dangers of hypnosis, 98–101
  Hynek's experiments, 104–106
  legal aspects, 91, 96, 100
  problem of contact, 94–98
hypotheses, 195–213
  of civilizations on other planets, 131–132
  earthbound aliens, 195–196, 200–201, 203–204
  extraterrestrial, 14, 54–55, 195, 196–200
  genetic programming, 195–196
  interlocking universes, 212–213
  possible scenarios, 191–195
  psychic, 61, 207
  psychic projection, 209–210
  rejected, 23, 51–52
  secret human base, 196, 205–209

I

Identified Flying Objects (IFOs), 3, 13, 27, 206
Indian cultures, and sky people, 119–120, 200
Industrial Revolution, and disappearance of "Lutins," 201
inoculation technique, 55
intelligent life on other planets, belief in, 131, 229–230, 233–234
interlocking universes, 212–213

Index **243**

interviews with Allen Hynek
  caliber of witnesses,
    143–145
  encounters with occupants,
    85–86, 90–99, 103–106
  hypotheses, 191–213
  investigation of Ely case,
    37–45
  physical natures and
    psychology, 118–123,
    126–136
  publications, 180–190
  repeaters, 53–57
  research evolution,
    153–177
  research on UFOs, 23–37
  research progress, 58–62
  scientific topics, 64–82
interviews with Arthur
  Hastings
  caliber of witnesses,
    144–147
  encounters with occupants,
    94–103
  hypotheses, 192–212
  physical natures and
    psychology, 118–122,
    126–135
  publications, 179–190
  repeaters, 53–57
  research evolution,
    154–176
  research on UFOs, 22–37
  research progress, 58–62
  scientific topics, 64–82
interviews with Jacques Vallée
  caliber of witnesses,
    144–148
  encounters with occupants,
    83–84, 90–104
  hypotheses, 191–213
  investigation of Ely case,
    37–44
  physical natures and
    psychology, 119–135
  publications, 179–190
  repeaters, 53–57
  research evolution,
    153–177
  research on UFOs, 23–37
  research progress, 58–62
  scientific topics, 64–82
investigation, 73–82
  catagories of, 27
  computer, 73–78
  magnetic, 58, 81–82

new methodology
  needed, 18
  photographic, 58, 78–81
  questionnaires, 103–104
  See also France; research
Invisible College, 153–156, 168
  The Invisible College
    (Vallée), 199
"Iron Mountain Report,"
  132, 135

J

Jacobs, David, *The UFO
  Controversy in America,*
  148n, 187
Japanese UFO groups, 184
jealous phenomenon, 66
Jeans, Sir James, 201

K

Kachina dolls, 119
Keel, John
  Mothman Prophecies, 188
  Operation Trojan Horse, 188
Keyhoe, Donald, 134, 158,
  162, 185
  Flying Saucers Are Real, 185
  "Project Lure," 195
Klass, Philip, 188
Koestler, Arthur, "Twilight
  Bar," 191
Kreskin, "Amazing," 105–106
Kuettner, Joachim, "UFOs,
  An Appraisal of the
  Problem," 173n

L

landings, 14, 17, 26
  Connersville, 24
  French investigations of, 58
  Saskatchewan, Canada,
    215–224
LeDonne, Robert, 225–231
Leningrad, USSR, 235–236
levitation, 19, 19n
light beam at car,
  Gleichen, 19n
lightning calculators, 210–211
lights on/from UFOs, 14
  astronomer sightings, 32n
  Australia, RAF, 59
  Boianai, 48–50
  Connersville case, 24
  Dexter sighting, 163–164
  Ely case, 19–21, 39–43
  Hill, 87

as photographic
  evidence, 80
Woodcliff, New Jersey,
  226–227, 228, 230
literature. *See* publications
Little People, 121–122, 200
Long Island, New York, 32n
Lorenzen, James and Coral,
  Shadow of the Unknown, 188
Lovell, Jim, 64
Low, Robert, 174
Lumieres dans la Nuit, 183
luminosity, 10, 14
  Australia RAF, 59
  Boianai, 50
  northern US Plains, 108,
    111–112
  Torino, 60
  See also lights on/from
    UFOs
Lunan, Duncan, 197
"Lutins," disappearance of, 201

M

magnesium sample from
  Ubatuba, 67
magnetic detection, 58, 81–82
Manchester, Georgia, 27n
Manhattan Project, 205–206
Mannor, Frank, 163
Marley, Lt., 169
Mauna Kea, Hawaii, 32n
McCampbell, James, 28
McDivitt, Jim, 64
McDonald, James, vii–viii, 157,
  161, 166
Mekis, Karl, 150–151
Men in Black, 67–68
  Damon, 68
  northern US Plains,
    115–117, 135
Menzel, Donald
  as critic, 23, 33, 51
  and Invisible College,
    153–155
  natural causes, as
    explanations, 56–57
  skeptical books, 187, 188
mesmerizing, 62
meteors, 15, 27, 35
Meu creatures, 123–126
Michel, Aimé, 4, 54, 160,
  167, 168
  *Flying Saucers and the
    Straight-Line Mystery,*
    167n, 195, 207
  the nature of contact, 199

Miller, Emmett, 99
Monroe, Robert, *Journeys Out of The Body*, 95
Moody, Sgt, 164, 168
Morier, Ron, 221, 222–224
Moseley, James, 146n
Mount Palomar, 145, 147
Mutual UFO Network (MUFON), 75, 104, 183
myth
    function of, 189–190
    little people, across the world, 200–201

N

NASA (National Aeronautics and Space Agency)
    astronaut sightings, 63–65
    contingency plan for encounters, 130
    hypothesis of future, 197, 198, 199, 201
Nevada case in Ely, 18–22, 31, 37–45, 194, 213
New Guinea case, 47–53, 56–57, 62, 153
New Jersey sighting, 225–231
New York
    Long Island, 32n
    Walesview, 134–135
New York Times, 134, 135
NICAP (National Investigation Committee on Aerial Phenomenon)
    failure to publish data, 160, 162
    Hynek's view of, 158–159
    information on sightings, 73, 75–76
    membership and publications, 181–183
    possible report, 175
nonsense, UFOs as, 31–33, 37
northern US Plains case, 107–117
northern US states, Capella case, 140–143
nuclear energy, 36–37, 131

O

occult connections with UFOs, 120–126, 208–209
occupants of UFOs
    avoidance of people, 14, 15, 62, 194
    biology, forms and shapes of, 203–204

Boianai, 49–53
    cultural reactions to, 94
    Eagle River, 122, 128
    Hill, 84, 87–89, 93
    lack of purpose, 31
    northern US Plains, 108–110, 112–114, 122
    Pascagoula, 97
    repairing their equipment, 54
    statistics on, 28
    study of, 4–5
    waving back to witnesses, 52–53
out-of-the-body experiences, 212–213

P

Paciello, James, 32n
Page, Thornton, 156
Parker, Calvin, 86, 95, 97
    See also Pascagoula, Mississippi
Pascagoula, Mississippi
    account of, in book, 187
    close encounter, 83, 86, 92, 93, 94–95, 96–97, 99
    Rolling Stone article on, 95
Passport to Magonia (Vallée), 4, 13n, 75, 180
peace, hope for, with space race proposals, 132
Phillips, Ted, 24, 28, 29, 210
    Physical Traces Associated With UFO Sightings, 24n
    Saskatchewan encounter, 215, 217, 218–224
photographic evidence, 53, 57, 58, 78–81
    confiscated by Men in Black, 67–68
    fake pictures of hoaxes, 146–147, 148
    ignored by astrophysicists, 65
physical effects. *See* UFO effects: physical
physics, 36
Poher, Claude, 15, 28, 29, 81–82
police, taking reports seriously, 221, 222–224, 227–228
Portugal, research, 26
Powers, William, 141, 142, 163
Project Blue Book
    beginnings of, 5, 154n

    under Captain Hardin, 155–156, 159, 169–170
    closing of, 2, 226, 230
    Congressional hearings, 60, 158
    fallacy with statistics, 159–161
    files, compared with French cases, 167
    Hynek as consultant on, 155–171
    and NICAP, 158–159
    UFO statistics from, 27
Project Grudge, 5, 155
Project Sign, 5, 27, 143, 155
psychic aspects, 61–62
psychic events
    Ely, 21–22, 44–45
    paranormal, 4, 6, 12, 34, 209–210
psychic projection, 209–210
public opinion
    belief in UFOs correlated with education and income, 13, 31
    changing, on UFOs, 161–162, 175–176
    education on research, 23, 34–35
    needing an explanation, 33
    sightings by, and awareness of Americans, 233–234
    survey in Detroit, 193
    See also Gallup Poll
publications, 179–190
    amateur UFO groups, 181–185
    author's books, 179–181
    myths, 189–190
    reading the Condon Report, 181
    recommended books, 185–187
    skeptical books, 188–189

Q

questionnaires, 103–104
Quintanilla, Col., 164, 169

R

radar, 7, 14
    Air Force report of Hills' encounter, 84
    Australian RAF, 59–60
    confirmations, 60, 135

Index   **245**

as example of development
of scientific discovery,
239
military lack of interest in,
65–66
Torino, 60
radio communications, 198
"Report From Iron Mountain,"
132, 135
reports. *See* UFO reports
research, 22–37
computer, 73–78
establishment of scientific
research, 23
laboratory analysis, 53–54
scientific study of UFOs,
16, 18, 35, 53–54,
153–156
technological spinoffs, 69
who should study UFOs?,
68–70
worldwide, 4, 23, 58–59
See also investigation;
scientific study
of UFOs
Rhine, J. B., 167–168
ridicule
"flying saucers" as, 166,
171–172
See also fear of ridicule
Roach, Franklin E., 65
Robertson, H.P., 71, 156, 166
Robertson Panel, 71, 72,
155–156
Rome, New York, 135
Royal Canadian Mounted
Police (RCMP), 215, 221,
222–224
Ruppelt, Edward J., 5, 154, 155,
162, 189

S

Sagan, Carl, 132, 196, 198
Saskatchewan, Canada, 215–224
satellite tracking program,
65, 158
Saunders, David, 28, 74, 173, 174
UFOs? Yes!, 173n, 174n, 187
scenarios of future UFO
developments, 191–195
Schirra, Walter, 64
Schwartz, Ed, 83–94
science
changing viewpoints in
science, 33–37, 166, 176
education on research, 23,
34–35

imagining the future, 11,
12, 36
mysteries and discoveries,
31–33, 34–37, 61, 136
predictions of the future,
201–202
public attitudes on, 16,
18, 31–33, 35, 136, 144,
175–176
publication of data,
185–186
radar as example of
development of scientific
discovery, 239
scientific debate about UFOs
Blue Book fallacy with
statistics, 159–161
change of climate over the
years, 161–162
Condon Committee,
behind the scenes,
172–175
credibility gap, 175–177
cultism, 171–172
European investigations,
166–168
failure of Air Force
"investigations," 156–159,
168–171
Invisible College,
153–156, 168
looking at data, 164–166
swamp gas incident,
162–164
scientific study of UFOs, 16, 18,
32, 35, 53–54
changes with Invisible
College, 153–156, 168
See also research
scientists
astronaut sightings, 63–65
credibility gap with the
public, 189
government cover-up,
70–73
interest in studying UFOs,
18, 23, 93, 162, 184
lack of interest in studying
UFOs, 143–144, 154
solving mysteries and
puzzles, 32–33, 35–37
who should study UFOs?,
68–70
Shaw, Bernard, 166
sighting durations, statistics
on, 29
sighting times, 15–16, 17, 26

Simak, Clifford, *Out of Their
Minds*, 193
Simon, Benjamin, 84–85,
88–89, 90, 97–98, 99
Simonton, Joe, 127–129
skeptics, 15, 31, 71–72
books by, 188–189
Hynek as, 4, 5, 72
occult traditions, 121
sketches as evidence, 81
sky people in Indian cultures,
119–120, 200
sociological perspective of
UFOs, 182, 184, 185
sociological study of
witnesses, 144
Socorro case, 2
sounds from UFOs, 10
Chesterfield Inlet, 8
Ely, 42–43
Saskatchewan, 216, 221
Woodcliff, New Jersey,
227, 230
space. *See* astronomy
Spain
composition of groups of
witnesses, 123, 144
research, 26, 28, 58, 76
speed of light, 34
Spielberg, Steven, 2
Spitzbergen, Norway, 67
Strange Phenomena, 188–189
Stranges, Frank, *The Stranger at
the Pentagon*, 187
Struve, Otto, 32n, 130, 201
Sturrock, Peter, 173
swamp gas, as explanation
of UFOs, 2, 4,
162–164, 172

T

technological-cultural
gap, 237
technology from UFOs,
68–69, 73
Tesla, Nicola, 208
Texas
Aurora, 148–149
Damon, 68
time loss, of witnesses
Hill, 84, 88, 118
northern US Plains,
114–115, 118
Torino, Italy, 60
trajectories, 14, 15, 206–207
Dzhuze, 235
Leningrad, 235–236

246 THE EDGE OF REALITY

trajectories *(continued)*
  Saskatchewan, 216,
    218–219
  See also flight and
    movement of UFOs

U

Ubatuba, Brazil, 67
UFO (unidentified flying
  object), definition of, between
  mythological objects, 177
UFO characteristics
  absurdity, 54–56, 129
  dematerialization, 37
  game theory, 54
  hostility from, 133–136
  inoculation, 55
  intelligent control, 14, 32n,
    54, 62, 63
  intentionality, 14, 53, 54,
    194–195
  lack of purpose, 31
  repeaters, 53–57
  space-time localization,
    14–15, 66
  time distribution, 15–16,
    17, 26
  See also craft (size and
    shape); flight and
    movement of UFOs;
    lights on/from UFOs;
    luminosity; occupants
    of UFOs; sounds from
    UFOs; trajectories
UFO effects: physical, 14,
  209–210
  electromagnetic effects, 14
  evidence in environment,
    24, 25, 53
  evidence in Saskatchewan
    field, 215–217, 220,
    221, 223
  hardware from crashes,
    66–67
  heat wave in jet fighter, 135
  melting of asphalt,
    133–134
  See also vehicle
    interferences
UFO effects: physiological, 14,
  25, 213
  healing of wound, 68
  hostile injuries, 133
  medical injuries, 69
UFO effects: psychological, 118
  paranormal aspects, 4, 6,
    12, 34, 209–210

  playing games with
    witnesses, 20–21,
    37–38, 43
  psychic aspects and events.
    See psychic aspects;
    psychic events
*The UFO Experience: A Scientific
  Inquiry* (Hynek), 6, 13n,
  180–181
UFO literature. *See* publications
UFO organizations, 181–185
  French, 183, 185
  Japanese, 184
  publications by, 181–185
  See also APRO; Center
    for UFO Studies;
    Mutual UFO Network
    (MUFON); NICAP
UFO phenomenon
  a jealous phenomenon, 66
  popular misconceptions,
    24, 61
  problem: statement in six
    points, 11–18
  real but unknown, 4, 28, 35
  worldwide aspects, 3, 9,
    13, 23, 25, 28, 53, 55–56,
    58–60, 167
UFO reports
  consistency of, 12, 13–16,
    28, 59, 194
  existence of reports, 3–4,
    12–13, 24
  number of, 3, 30, 157,
    206–207
  patterns, 10, 13, 25
  related to population
    density, 15, 17, 25, 233
  times of sightings, 15–16,
    17, 26
  unreported, 3, 16,
    65–66, 206
  to Vice President
    Humphrey, 175
  See also Condon Report;
    Project Blue Book
UFO statistics, 26–30
  age distribution of
    witnesses, 30
  close encounters, 26, 28–30
  correlation with magnetic
    fields, 82
  distance in reports, 29
  duration, 29
  French data, 160, 168
  landings, 17, 28, 180
  number of reports, 30
  occupants, 28

  population density, 15, 17
  unexplained reports, 27
UFO waves, 9
  in China, 55
  in France in 1954, 4, 30,
    82, 167
  in Spain, 144, 192
  in US in 1896, 148n
UFOCAT, 74, 173
United Nations, 92
United States Air Force
  Blue Book fallacy with
    statistics, 159–161
  Capella case, 140–143
  Clearwater, Florida
    experiment, 137–140
  confiscation of evidence by
    Men in Black, 67–68
  data, filing system of
    reports, 73
  data files, and patterns of
    data, 194
  in Eagle River incident, 128
  effect on amateur
    groups, 181
  failure in informing public,
    34–35
  failure of "investigations,"
    156–159, 168–171
  Hynek's book, 181
  Hynek's research on UFO
    reports, vii–viii, 1–2,
    4–5, 143
  hypothesis of secret
    experimental craft, 205
  jet fighter mechanical
    malfunction, 134
  in northern US Plains
    incident, 115
  possible future scenario, 193
  [possible] governmental
    cover-up, 70–72, 185
  questionnaire for
    witnesses, 104
  radar, denial of
    confirmations, 60
  radar report of Hills'
    encounter, 84
  sightings no longer
    recorded, 226, 230
  See also Condon
    Committee; Condon
    Report; Project Blue
    Book; Project Grudge;
    Project Sign
universes, multiple, 212–213
University of Colorado, and
  Condon Report, 174

U.S. Pentagon, orders to debunk UFO reports, 155–156, 161
USAF. *See* United States Air Force
USSR
  cosmonauts as witnesses, 64
  sightings by pilot and scientist, 235–236
  UFO reports in, 55
  UFO studies, 58, 92

V

Vallée, Jacques
  Anatomy of a Phenomenon (Vallée), 4, 13n, 121–122, 179–180
  Challenge to Science (Vallée), 4, 13n, 26n, 173, 180
  conversations with Paul Hynek, vii–viii
  The Invisible College, 199
  Passport to Magonia (Vallée), 4, 13n, 75, 180
  research on UFOs, 4–6, 17, 26–27, 29

research with Allen Hynek, 1–2
  See also interviews with Jacques Vallée
vehicle interferences, 10, 14
  Bananeiras, 19n
  Chesterfield Inlet, 8
  Ely case, 19–20, 22, 38–39, 41
  Gleichen, 19n
  Hill, 88, 89
Venus, 49, 51–52
von Daniken, Erich, 189–190, 191

W

Walesview, New York, 134–135
Walton, Joe, 63
Weber-Richter, Frank, 150–151
Whipple, Mr., 155, 158
White, Robert, 63
witnesses of UFOs
  accuracy of reports, in experiment, 137–140
  age distribution, 30, 144
  astronauts as, 63–65, 130

caliber of, 13, 143–145
  correlation of education and income, 13, 31
  credibility of, 12, 24, 47, 164–165, 223
  fear of ridicule, 3, 12
  number of, 29
  scientists as, 32n, 57, 235–236
  single witness cases, 128–129
  See also animal reaction; contactees; fear of witnesses; hostility from UFOs; hostility from witnesses toward UFOs; time loss
Wright-Patterson AFB, 5, 154–155, 157, 161, 164, 165

Y

Young, John, 63, 64

Z

Zamora, Lonnie, vii

TO OUR READERS

MUFON BOOKS

The mission of MUFON BOOKS, an imprint of Red Wheel Weiser, is to publish reasoned and credible thought by recognized authorities; authorities who specialize in exploring the outer limits of the universe and the possibilities of life beyond our planet.

ABOUT MUFON

The Mutual UFO Network (*www.MUFON.com*) was formed by concerned scientists and academic researchers in 1969 for the specific purpose of applying scientific methods to the serious study of UFO sightings and reported human/alien interactions. MUFON's mission is "The Scientific Study of UFOs for the Benefit of Humanity" with the intent to unveil and disclose credible information free of distortion, censorship, and lies, and prepare the public for possible implications.

ABOUT RED WHEEL/WEISER

Red Wheel/Weiser (*www.redwheelweiser.com*) specializes in "Books to Live By" for seekers, believers, and practitioners. We publish in the areas of lifestyle, body/mind/spirit, and alternative thought across our imprints, including Weiser Books, and Career Press.